D1747398

Articulations between Tangible Space,
Graphical Space and Geometrical Space

Education Set

coordinated by
Angela Barthes and Anne-Laure Le Guern

Volume 14

Articulations between Tangible Space, Graphical Space and Geometrical Space

Resources, Practices and Training

Edited by
Claire Guille-Biel Winder
Teresa Assude

iSTE WILEY

First published 2023 in Great Britain and the United States by ISTE Ltd and John Wiley & Sons, Inc.

Apart from any fair dealing for the purposes of research or private study, or criticism or review, as permitted under the Copyright, Designs and Patents Act 1988, this publication may only be reproduced, stored or transmitted, in any form or by any means, with the prior permission in writing of the publishers, or in the case of reprographic reproduction in accordance with the terms and licenses issued by the CLA. Enquiries concerning reproduction outside these terms should be sent to the publishers at the undermentioned address:

ISTE Ltd
27-37 St George's Road
London SW19 4EU
UK

www.iste.co.uk

John Wiley & Sons, Inc.
111 River Street
Hoboken, NJ 07030
USA

www.wiley.com

© ISTE Ltd 2023

The rights of Claire Guille-Biel Winder and Teresa Assude to be identified as the authors of this work have been asserted by them in accordance with the Copyright, Designs and Patents Act 1988.

Any opinions, findings, and conclusions or recommendations expressed in this material are those of the author(s), contributor(s) or editor(s) and do not necessarily reflect the views of ISTE Group.

Library of Congress Control Number: 2022950715

British Library Cataloguing-in-Publication Data
A CIP record for this book is available from the British Library
ISBN 978-1-78630-840-5

Contents

Preface xv
Claire GUILLE-BIEL WINDER and Teresa ASSUDE

**Part 1. Articulations between Tangible Space, Graphical Space
and Geometric Space** 1

**Chapter 1. The Geometry of Tracing, a Possible Link Between Geometric
Drawing and Euclid's Geometry?** 3
Anne-Cécile MATHÉ and Marie-Jeanne PERRIN-GLORIAN

   1.1. Introduction 3
   1.2. Geometry in middle school 5
      1.2.1. What underlying axiomatics? 5
      1.2.2. An example 6
      1.2.3. The current lack of consistency 8
   1.3. Geometry of tracing, a possible link between material geometry and Euclid's
   geometry? 8
      1.3.1. Figure visualization and figure restoration 9
      1.3.2. The geometrical use of tracing instruments, a first step to make sense to an
      axiomatic 10
      1.3.3. Distinguishing between the hypothesis and the conclusion 12
      1.3.4. Restoration, description, construction of figures and geometric language . 14
   1.4. Dialectics of action, formulation and validation with regards to the reproduction
   of figures with instruments. 15
      1.4.1. Formulation situations and possible variations 15
      1.4.2. Validation situations 17
   1.5. From tracing to the characterization of objects and geometric relationships . . 18
      1.5.1. On the concepts of segments, lines and points 18
      1.5.2. On the notion of perpendicular lines 21

1.6. Towards proof and validation situations in relation to figure restoration 27
    1.6.1. Equivalence between two construction programs and the need for proof. 27
    1.6.2. Validation situations involving programs for the construction of a square and introducing a proof process 29
1.7. Conclusion 31
1.8. References 32

Chapter 2. How to Operate the Didactic Variables of Figure Restoration Problems? 35
Karine VIÈQUE

2.1. Introduction 35
2.2. Theoretical framework 35
    2.2.1. Studying a specific type of problem: figure restoration 35
    2.2.2. Studying the concepts involved in figure restoration problems 37
2.3. Values of the didactic variables of the first problem family 39
    2.3.1. Values of the didactic variables for the "figure" and the "beginning of the figure" 39
    2.3.2. Value for the didactic variable "instruments made available" 40
    2.3.3. Rules of action and theorems-in-action associated with development on the geometrical usage of the ruler 41
2.4. Conclusion 44
2.5. References 44

Chapter 3. Early Geometric Learning in Kindergarten: Some Results from Collaborative Research 47
Valentina CELI

3.1. The emergence of the first questions. 47
3.2. Theoretical insights 48
    3.2.1. Global understanding and visual perception of geometric shapes 48
    3.2.2. Operative understanding and visual perception of geometric shapes. . . . 49
    3.2.3. Topological understanding and visual perception of geometric shapes . . 50
    3.2.4. Haptic perception 51
    3.2.5. Association of visual and haptic perceptions: towards a sequential understanding of geometric shapes 52
3.3. The role of language in early geometric learning 53
    3.3.1. But which lexicon? 54
    3.3.2. Verbal and gestural language 58
3.4. Assembling shapes 60
    3.4.1. Free assembly of shapes 60
    3.4.2. Assembling triangles 62
3.5. Gestures to learn 68

3.6. Conclusion	69
3.7. References	71

Chapter 4. Using Coding to Introduce Geometric Properties in Primary School 73
Sylvia COUTAT

4.1. Coding in geometry	73
4.2. Two examples of communication activities requiring the use of coding	75
4.2.1. A co-constructed coding	75
4.2.2. Personal coding	77
4.3. Conclusion: perspectives on the introduction of coding in geometry	78
4.4. References	79

Chapter 5. Freehand Drawing for Geometric Learning in Primary School 81
Céline VENDEIRA-MARÉCHAL

5.1. Introduction	81
5.2. Drawings in geometry and their functions	82
5.3. Freehand drawing in research	83
5.4. Exploring the milieu around a freehand reproduction task of the Mitsubishi symbol on a blank white page	84
5.4.1. Freehand drawing reveals a reasoning between spatial knowledge and geometric knowledge	87
5.4.2. Freehand drawing as a dynamic process to build and transform knowledge	88
5.5. Conclusion	89
5.6. References	90

Part 2. Resources and Artifacts for Teaching 93

Chapter 6. Use of a Dynamic Geometry Environment to Work on the Relationships Between Three Spaces (Tangible, Graphical and Geometrical). 95
Teresa ASSUDE

6.1. Added value with a dynamic geometry environment: the ecological and economical point of view	95
6.2. Tangible space, graphical space and geometric space	100
6.3. Designing situations for first grade primary school	103
6.3.1. Our choices for designing situations	104
6.3.2. Presentation of situations	104
6.4. Analysis of the situations for the first-grade class	105
6.4.1. Instrumental dimension: perceptive–gestural level	105

6.4.2. Instrumental dimension: spatial–geometric relationships 106
6.4.3. Instrumental dimension: exploration and graphical space. 107
6.4.4. Instrumental dimension: tool-geometric space symbiosis. 108
6.4.5. Praxeological dimension 109
6.4.6. Praxeological dimension: observe and describe 111
6.5. Conclusion 113
6.6. References 115

Chapter 7. Robotics and Spatial Knowledge 119
Emilie MARI

7.1. Introduction 119
7.2. Theoretical framework and development for a categorization
of spatial tasks. 120
   7.2.1. Spatial knowledge 120
   7.2.2. Types of spatial tasks. 121
   7.2.3. Types of tasks and techniques 121
7.3. Research methodology 122
7.4. Analysis: reproducing an assembly 123
   7.4.1. Test item. 123
   7.4.2. Test results. 124
   7.4.3. Analysis of the results 125
7.5. Conclusion 126
7.6. References 127

**Chapter 8. Contribution of a Human Interaction Simulator to Teach
Geometry to Dyspraxic Pupils** 129
Fabien EMPRIN and Edith PETITFOUR

8.1. Introduction 129
8.2. General research framework 130
   8.2.1. Teaching geometry 130
   8.2.2. Dyspraxia and consequences for geometry 131
8.3. What alternatives are there for teaching geometry?. 132
   8.3.1. Using tools in a digital environment 132
   8.3.2. Dyadic work arrangement 135
8.4. Designing the human interaction simulator 138
   8.4.1. General considerations 138
   8.4.2. Choice of instrumented actions 139
   8.4.3. Interaction choices 140
   8.4.4. Ergonomic considerations 142
8.5. Initial experimental results 143
   8.5.1. Data collected 144
   8.5.2. Jim's diagnostic evaluation 144

8.5.3. Analysis of the first experimentation	146
8.5.4. Conclusion	150
8.6. References	152

Chapter 9. Research and Production of a Resource for Geometric Learning in First and Second Grade ... 155
Jacques DOUAIRE, Fabien EMPRIN and Henri-Claude ARGAUD

9.1. Presentation of the ERMEL team's research on spatial and geometric learning from preschool to second grade	155
9.1.1. Origins of the research	156
9.1.2. Introduction to the chapter	156
9.2. Learning to trace straight lines	157
9.2.1. Significance of the straight line	157
9.2.2. Initial hypotheses	157
9.2.3. The RAYURE situation	159
9.2.4. Using straight lines	160
9.2.5. A few summary elements	161
9.3. Plane and solid figures	162
9.3.1. Findings and assumptions	162
9.3.2. The SQUARE AND QUASI-SQUARE situation	163
9.3.3. The emergence of criteria for comparing solids: the IDENTIFYING A SOLID situation	165
9.3.4. Identification of cube properties: the CUBE AND QUASI-CUBE situation	166
9.3.5. Progression on solids and plane figures	167
9.4. The appropriation of research results by the resource	168
9.5. Conclusion	169
9.6. References	170

Chapter 10. Tool for Analyzing the Teaching of Geometry in Textbooks ... 171
Claire GUILLE-BIEL WINDER and Edith PETITFOUR

10.1. General framework and theoretical tools	172
10.1.1. Didactic co-determination scale, mathematical and didactic organizations	172
10.1.2. Reference MO and theoretical tools for analysis	174
10.2. Analysis criteria: definition and methodology	181
10.2.1. Institutional conformity	181
10.2.2. Educational adequacy	182
10.2.3. Didactic quality	182
10.3. Introducing the analysis grid	183
10.3.1. Analysis of tasks and task types	183

10.3.2. Analysis of techniques 184
10.3.3. Analysis of knowledge 185
10.3.4. Analysis of ostensives. 186
10.3.5. Analysis of organizational and planning elements 189
10.3.6. Summary. 191
10.4. Conclusion 191
10.5. References 192

Part 3. Teaching Practices and Training Issues. 197

Chapter 11. Study on Teacher Appropriation of a Geometry Education Resource. 199
Christine MANGIANTE-ORSOLA

11.1. Introduction. 199
11.2. Research background 200
11.2.1. Study on dissemination possibilities in ordinary education 200
11.2.2. Resource design approach 201
11.2.3. A working methodology based on assumptions 202
11.2.4. Designing a situation using the didactic engineering approach for development 205
11.3. Focus on the adaptability of this situation to ordinary education. 206
11.3.1. Details about the theoretical framework and the research question . . . 206
11.3.2. Presentation on the follow-up of teachers, details of the research question and the methodology 207
11.3.3. Presentation of the analysis methodology. 208
11.4. Elements of the analysis. 209
11.4.1. Analysis a priori of the situation and anticipatory analysis of the teacher's activity. 209
11.4.2. Analysis of practices 211
11.5. Conclusion 217
11.6. References 219

Chapter 12. Geometric Reasoning in Grades 4 to 6, the Teacher's Role: Methodological Overview and Results. 221
Sylvie BLANQUART

12.1. Introduction. 221
12.2. Theoretical choices and the problem statement 221
12.2.1. Geometrical paradigms 222
12.2.2. The different spaces. 223
12.2.3. Study on reasoning 223
12.2.4. The role of the teacher 225
12.2.5. Problem statement. 225

12.3. Methodology	225
12.3.1. General principle	225
12.3.2. The situations	226
12.3.3. Analysis methodology	226
12.4. Conclusion	227
12.5. References	229

Chapter 13. When the Teacher Uses Common Language Instead of Geometry Lexicon 231
Karine MILLON-FAURÉ, Catherine MENDONÇA DIAS, Céline BEAUGRAND and Christophe HACHE

13.1. Introduction	231
13.2. An attempt to categorize the uses of common vernacular terms in place of geometry lexicon terms within teacher discourse	232
13.2.1. The phenomenon of didactic reticence	232
13.2.2. The phenomenon of semantic analogy: comparison with common concepts to construct meaning for mathematical knowledge	233
13.2.3. The phenomenon of lexical competition: use of common vernacular terms to designate common concepts	234
13.2.4. The phenomena of repeating pupil formulations	235
13.2.5. The phenomenon of didactic repression	236
13.3. Conclusion	237
13.4. References	238

Chapter 14. The Development of Spatial Knowledge at School and in Teacher Training: A Case Study on *1, 2, 3... imagine!* 241
Patricia MARCHAND and Caroline BISSON

14.1. Introduction and research question	241
14.2. Conceptual framework	243
14.2.1. Components set to address SK in primary school	244
14.2.2. Levels of abstraction that value SK	245
14.2.3. Main variables in situations where SK is valued	246
14.3. Presentation of the activity 1, 2, 3 ... imagine!	247
14.4. Experiments with this activity in primary school and in teacher training in Quebec	251
14.4.1. Teaching sequence experimented in primary school	251
14.4.2. Teaching sequence tested in teacher training	254
14.5. Experiment results	255
14.5.1. Experiment results of the teaching sequence in primary school	255
14.5.2. Experiment results of this teaching sequence in teacher training	257
14.6. Conclusion	259
14.7. References	260

Chapter 15. What Use of Analysis a priori by Pre-Service Teachers in Space Structuring Activities? 265
Ismaïl MILI

    15.1. Introduction – an institutional challenge of transposing
didactic knowledge 265
        15.1.1. Choice of external transposition: institutional constraints 265
    15.2. Theoretical framework 267
        15.2.1. Choice of internal transposition: the moments of the study
of the analysis a priori 268
    15.3. Research questions 269
    15.4. Methodology 269
        15.4.1. Selection of activities and brief analysis 270
    15.5. Results 272
    15.6. Conclusion 273
    15.7. References 273

Part 4. Conclusion and Implications 275

Chapter 16. Questions about the Graphic Space: What Objects? Which Operations? 277
Teresa ASSUDE

    16.1. Semiotic tools of geometric work and graphic space 277
    16.2. Graphic space: graphic expressions, denotation and meaning 280
        16.2.1. How can we define the graphic space? 280
        16.2.2. Which objects in the graphic space? 280
        16.2.3. Graphic expressions: which operations? 282
    16.3. References 285

Chapter 17. Towards New Questions in Geometry Didactics 289
Claire GUILLE-BIEL WINDER and Catherine HOUDEMENT

    17.1. Current questions in geometry didactics 289
    17.2. Continuities and breaks in the teaching of geometry 291
        17.2.1. Institutional continuity? 291
        17.2.2. Theoretical continuity from "geometry of tracing" to
"abstract geometry"? 291
        17.2.3. Praxis continuity from the "geometry of tracing" to
"abstract geometry" 294
    17.3. Articulation between resources, practices and teacher training 297
    17.4. References 299

Appendices	303
Appendix 1	305
Appendix 2	309
Appendix 3	311
Appendix 4	313
List of Authors	315
Index	317

Preface

Geometry is one of the oldest branches of mathematics. According to Brousseau, it "appears, through its aims, through its statements, through its methods, and through its multiple representations across the many branches of Mathematics and Science, sometimes in unexpected ways" (2000, p. 67, *translated by author*). Brousseau identifies in the teaching of geometry, on the one hand, a didactic means for "introducing mathematics", in that it offers, contrary to other domains, the possibility for teachers "to elicit in their students an activity recognized as authentically mathematical by most mathematicians themselves" (ibid.), and on the other hand, as a means by which to represent space. Geometry thus appears to be an area within teaching that is still relevant today. The report of the commission on the teaching of mathematics, directed by Jean-Pierre Kahane, goes a little further and highlights four advantages to teaching geometry (Kahane 2002): to appropriate a vision of space and its representations; to learn geometrical reasoning; to be initiated into aesthetic and cultural aspects; and to have access to certain acumen that is useful across many trades. Numerous works realized over these last few years testify to the importance that the didacticians of mathematics accord with it in compulsory education.

The teaching of geometry in elementary school refers to two fields of knowledge that are intimately linked, but not to be confused (Berthelot and Salin 1993): spatial knowledge that allows us to control our relationship to the surrounding space; and geometric knowledge that allows us to solve problems involving objects in the physical, graphical or geometric space. Due to the importance of these two fields, one of the questions addressed in this book is: How is this knowledge taken into

---

Preface written by Claire GUILLE-BIEL WINDER and Teresa ASSUDE.

account and/or articulated in the teaching of space and geometry in compulsory education, in teaching resources (curricula, textbooks, etc.), or in current teacher training?

Moreover, the teaching of geometry in elementary school is often associated with the manipulation of instruments. However, teaching cannot be limited to a game with tangible objects, but must allow for mediation with the world of theoretical objects. This semiotic mediation is at the heart of mathematical activity, and geometry is a domain wherein the question of this mediation inevitably arises. Moreover, a geometric activity brings into play the language register and the graphical register (in particular that of figures), which must be articulated (Duval 2005). The language activity is also coordinated alongside a physical activity, for figures that must be traced or modified, with instruments or by freehand. Gestures also have a special place in the geometric activity itself. Therefore, the second line of questioning is as follows: How to take into account this semiotic dimension (language, gestures, signs, etc.) of geometric activity? What is the role of artifacts (digital or tangible) in the teaching and learning of geometry?

This book aims to present some of the latest research in the didactics of space and geometry, to deepen some theoretical questions and to open up new reflections for discourse, as much on the approach of geometry itself and its connection with the structuring of space, as on practices within the classroom, the dissemination of resources, the use of different artifacts or the training of teachers on this subject. It mobilizes about 15 contributions from French-speaking researchers based in different parts of the world (France, Switzerland, Quebec). It is organized into three main parts, which we present in the following.

Part 1 deals with the articulations between tangible space, graphic space and geometrical space. The contribution of Anne-Cécile Mathé and Marie-Jeanne Perrin-Glorian explores possible continuities between physical geometry and theoretical geometry through a figure reproduction task for pupils at the end of French elementary school (9–11-year-olds). Continuing on from this, Karine Vièque's text deals with the question of choosing the values for the didactic variables (geometrical and physical) in the elaboration of shape reproduction problems at the beginning of French elementary school (6–8-year-old students), in order to develop different understandings of plane shapes, in terms of visualization and deconstruction. Valentina Celi focuses on early geometric learning in kindergarten (3–6-year-olds). She presents the first results of an ongoing research-action, in which problems concerning geometric shapes are tested, from manipulation to graphical tracing, from the global understanding of shapes to a more analytical understanding of them, by articulating visual and haptic modalities and by progressively introducing an appropriate lexicon. Sylvia Coutat's work deals with the representation of geometric properties in the graphical register by proposing avenues for reflecting through the introduction of coding geometric properties in elementary school. Finally, Céline Vendeira-Maréchal questions the impact and relevance of using freehand drawing in construction/reproduction tasks with 8- to 10-year-olds, in particular, to free them from the manipulative constraints implied by geometric instruments.

Part 2 is devoted to resources and artifacts for teaching. Three chapters deal with digital resources, two others with teaching proposals. Teresa Assude shows how a dynamic geometry software can be a tool through which to work on the relations between tangible space, graphical space and geometric space. Emilie Mari's contribution deals with the impact of using programmable floor robots on the development of spatial and geometrical knowledge in French elementary school, among 6–8-year-old students. Fabien Emprin and Edith Petitfour present, within the framework of instrument construction problems, a possible exploitation for dyspraxic pupils and the possibilities offered by a human interaction simulator for geometric learning. Concerning teaching resources, Jacques Douaire, Fabien Emprin and Henri-Claude Argaud propose a presentation on the evolution of the research questions by the ERMEL team, concerning the drawing of straight lines and discovering the characteristics of plane shapes and solids, by explaining an analysis of the knowledge of 6- to 8-year-old pupils as well as the proposed problems. Elements on the implementation of situations as well as on the progression and structuring of the resource are also discussed. Finally, Claire Guille-Biel Winder and Edith Petitfour try to determine what could enlighten, from a didactic point of view, the choice of textbooks in the framework of geometry teaching, by analyzing the proposals for teaching the notions of perpendicularity and parallelism in fourth grade (9–10-year-olds).

Part 3 of the book focuses on teaching practices and training issues. At the intersection between resource development, practice and training, Christine Mangiante-Orsola studies the process of appropriation, by three teachers, of a situation developed during a research project on the teaching of geometry from third grade to fifth grade (8–11-year-olds), and then questions the possibilities for enriching teaching practices. Next, two contributions focus on teaching practices. Sylvie Blanquart conducts a clinical analysis of the same sequence of shape reproduction in fifth grade (10–11-year-olds) and in sixth grade (11–12-year-olds), with the aim of identifying how, in the progress of their teaching project in plane geometry, teachers integrate (or not) the valid or erroneous reasoning implemented explicitly or implicitly by the students. Karine Millon-Fauré, Catherine Mendonça Dias, Céline Beaugrand and Christophe Hache are interested in the discourse of the mathematics teacher, when they employ a term used by vernacular language instead of the appropriate term from the geometry lexicon. In their contribution, Patricia Marchand and Caroline Bisson describe and analyze a teaching sequence aimed at developing spatial knowledge that is deployed over the first three cycles of schooling. They then address the issues of initial training, since this same sequence was adapted and experimented with students in training, with both mathematical and didactic objectives. Finally, Ismaïl Mili is interested in the professional knowledge mobilized by teachers in training, and provides an example concerning the mobilization of a priori analysis in the implementation of a space structuring activity.

January 2023

References

Berthelot, R. and Salin, M.-H. (1993). L'enseignement de la géométrie à l'école primaire. *Grand N*, 53, 39–56.

Brousseau, G. (2000). Les propriétés didactiques de la géométrie élémentaire : l'étude de l'espace et de la géométrie. *Actes du séminaire de didactique des mathématiques*, Rethymnon, 67–83, hal-00515110.

Duval, R. (2005). Les conditions cognitives de l'apprentissage de la géométrie : développement de la visualisation, différenciation des raisonnements et coordination de leurs fonctionnements. *Annales de didactique et de sciences cognitives*, 10, 5–53.

Kahane, J.-P. (ed.) (2002). *L'enseignement des sciences mathématiques : commission de réflexion sur l'enseignement des mathématiques*. CNDP, Odile Jacob, Paris.

PART 1

Articulations between Tangible Space, Graphical Space and Geometric Space

1

The Geometry of Tracing, a Possible Link Between Geometric Drawing and Euclid's Geometry?

1.1. Introduction

The aims for teaching geometry are multiple: to reason about spatial situations, to practice geometric drawing with instruments, to solve problems related to such drawings and to approach a theoretical model of space which allows for the modeling of objects or problems of tangible space, but which is above all, in French secondary education, a gateway to hypothetico-deductive reasoning.

Numerous didactic works have pointed out the break between the geometry of elementary school, where the properties of figures are produced and verified with instruments, and the proof-based geometry taught in secondary school. Deductive geometry consists of the study of problems concerning idealized objects, as defined by language, and the modes for validating the properties of shapes, which stimulates hypothetico-deductive reasoning. Houdement and Kuzniak (2006) have spoken of two paradigms, GI and GII, to describe these two relationships within geometry.

Following on Berthelot and Salin (1992) and Brousseau (2000), we have distinguished between a material geometry, whose aim is practical, and a theoretical geometry, which aims at a consistent model of space. However, the objects of theoretical geometry are not unrelated to those of material geometry. What are the links between objects and practices for each of these relationships with geometry? How can we think of a theoretical geometry in middle school that is based on the geometry of primary school? Our work on this theme for the last 20 years (Mathé

Chapter written by Anne-Cécile MATHÉ and Marie-Jeanne PERRIN-GLORIAN.

et al. 2020) is through the perspective of the search for a coherent teaching of geometry throughout compulsory education. This concern leads us to explore possible continuities between material geometry and theoretical geometry.

We are particularly interested in the graphical space of physical 2D drawings: it is a specific part of the tangible space, which seems to us to be able to play an interface role between the tangible 3D space, the framework of material geometry and the theoretical geometrical space. Indeed, material geometry, as well as theoretical geometry, place at the heart of their problems questions of construction and analysis of figures. Graphical space serves as a field of experimentation for theoretical geometry. The same object within the graphical space can be studied as a physical object drawn on a sheet of paper or a computer screen, but it can also represent relations between objects of theoretical geometrical space, or objects or a situation within tangible space (a case that we will not consider here).

Graphical objects are thus susceptible to different interpretations and sources of misunderstanding: is the physical geometric figure studied for its own sake or does it represent a theoretical geometric object? Moreover, the analysis of pupils' difficulties in geometry in middle school shows that a portion of these comes from the lack of flexibility in the way they look at physical geometric figures. The analysis of physical geometric figures involves cognitive issues related to the visualization of figures (Duval 2005), and these are intimately linked to geometric knowledge. The objects of theoretical geometry are related by statements that translate into visual or instrumental features of physical geometric figures. The understanding of definitions and theorems and their use in demonstration require a dimensional deconstruction of the figures (identifying figurative subunits of 1- or 0-dimension, in a figurative 2D unit). The pupils' toolbox of theorems and definitions is not neutral with respect to the cognitive abilities needed to analyze geometric figures, an example of which we will see later on.

It seems that the question of continuity in geometric learning throughout compulsory education must be tackled from both ends. What are the foundations of geometry in middle school that would allow teachers to build on what pupils have already learned through drawing geometric figures with instruments? How to work on tracing figures with instruments in elementary school in order to establish a relationship with these material figures that will encourage pupils to acquire the flexibility needed to make proofs in geometry? How to make the problems posed to pupils regarding the construction of figures with instruments evolve so as to progressively make the object of study evolve, from material figures to geometric figures defined by language, and to make the avenue of questioning being explored evolve towards approaches that require the characterization of objects and proofs?

In this chapter, we will first focus on the objects and practices of middle school geometry, which, we will show, depend on the underlying axiomatics. We will then define what we have called the *geometry of tracing*, which corresponds to a kind of practical axiomatics based on the use of tracing instruments (excluding measuring instruments) around a fundamental situation, the reproduction of a geometric figure, seen as an action situation. We will then show, with the aid of examples situations experimented in class, how the dialectics of formulation and validation can be developed in the geometry of tracing, and lay down the foundations for thinking about possible mechanisms to promote continuity between material geometry and theoretical geometry.

1.2. Geometry in middle school

In middle school, one of the objectives of teaching geometry is as an introduction into demonstration. The break between a geometry of tracing with instruments and the deductive geometry (based on proofs) used to take place in the 7th grade (1960s), and then in the 8th grade (Laborde 1990). Today, this introduction into theoretical geometry is done progressively; however, the transition to geometric objects begins in the 6th grade (see the importance given in the 6th grade textbooks to the introduction of vocabulary and notations).

1.2.1. *What underlying axiomatics?*

In order to make proofs about theoretical objects, we need a repertoire of definitions and theorems which, for the sake of overall coherence, must be based on an axiomatic foundation: prime objects which are only defined by relations which link them, and which are posited as axioms.

Even though it is not a question of presenting an axiomatic theory to middle school pupils, at least for the teachers, the teaching should, in our opinion, be based on an axiomatic foundation with a sequence that should allow us to demonstrate all the results, even though we do not carry out all the demonstrations with the pupils.

For centuries, the foundation of geometry teaching was Euclid's axiomatics, reworked and completed by Clairaut in the 18th century, by Legendre and Lacroix at the end of the 18th century and the beginning of the 19th, and by Hadamard at the turn of the 20th century. Taken up at the very beginning of this geometry, the criteria for the congruence of triangles proved to be a powerful demonstration tool completed by the proportional segments theorem and the criteria for the similarity of triangles. In France, the reform of modern mathematics at the end of the 1960s rejected the congruence of triangles and wanted to instead base geometry on linear algebra at the high school level, building a rather complex definition of geometrical objects in middle school, quickly abandoned. In the 1980s, the teaching of geometry in middle school was based on the study of geometric transformations, which became the main demonstration tool. Annie Cousin-Fauconnet (1995) wrote an axiomatic based on the properties of orthogonal symmetry. However, the level of conceptualization for geometric objects and the dimensional deconstruction required by a demonstration vary according to the theoretical apparatus at one's disposal: demonstrations through transformations are more difficult than demonstrations through the congruence of triangles. A good discussion of this topic can be found in Groupe IREM de géométrie (2021); however, for our purposes, we will limit ourselves here to an example.

1.2.2. *An example*

What can we say about a triangle with two equal heights? We can first note that the construction of such a triangle is a problem in itself (see the discussion in Perrin and Godin 2018). We cannot construct the triangle from two heights without knowing more about these heights, which necessarily verify other properties. We can start by reasoning about a false figure (Figure 1.1(a)). Construction trials can make one think that the triangle is isosceles.

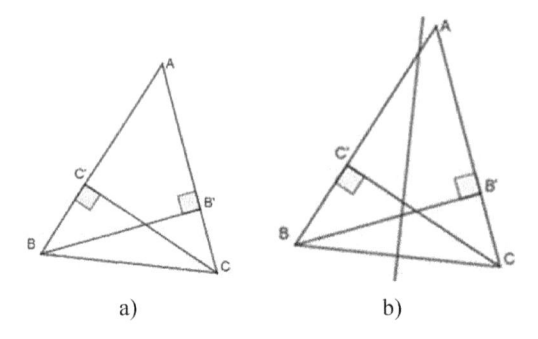

Figure 1.1. *A preliminary false figure*

This can be demonstrated in several ways using the congruence of triangles. For example, triangles ABB' and ACC' have the angle A in common and each a right angle; therefore, all their angles are equal. Moreover, BB' = CC'. They are congruent by the ASA criterion[1]. We deduce the equality of the homologous elements; thus, AB = AC. The only possible difficulty in the analysis of the

---

1 If two triangles have an equal side between angles equal to each, these triangles are congruent.

geometric figure is that the triangles overlap, but the elements to be considered are sides or angles, which is compatible with a vision of the figure as a superposition of surfaces whose edges and vertices must be considered: the sides are elements D1 on D2. To show that the triangle is isosceles using symmetry, we can show that the perpendicular bisector (d) of [BC] is the axis of symmetry for the figure, and to do this, we can show that B' and C' are symmetrical about this axis. The first difficulty, if we have constructed a figure that is slightly less false than this one (Figure 1.1(b)), consists of not using the fact that the axis passes through A, which is precisely what we want to demonstrate.

To show that B' and C' correspond to each other in the symmetry of axis (d), they must be seen as the intersection of lines (in this case circles) that correspond to each other in terms of the symmetry (Figure 1.2). The circle of diameter [BC] is invariant in terms of the symmetry: the points of this circle are transformed into other points along this circle. The circle with center B and radius BB' is transformed into a circle with center C and the same radius, passing through C', because BB' = CC'. B', the intersection of the first and second circles, is transformed into a point of intersection for the first and third circles, C'. The lines (BC') and (CB') are therefore symmetrical with respect to the axis, and their point of intersection A is on the axis of symmetry.

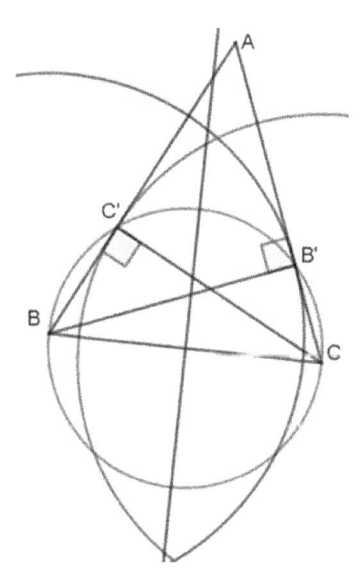

Figure 1.2. *A completed figure to solve with symmetry*

We can see the difficulty inherent to the demonstration: we have to see the point A as the intersection of two lines, themselves defined by two points, and to see these points themselves as the intersections of circles that we have to consider for this purpose, since they are not present in the initial statement. This requires a high level of dimensional deconstruction (Duval 2005).

This example is meant to show the complexity of a demonstration with transformations, but it is much simpler to demonstrate that the triangle ABC is isosceles by calculating its area in two ways. However, we must admit the existence of areas and think of using areas to show the equalities of lengths or angles!

1.2.3. *The current lack of consistency*

In the 1990s and 2000s, the study of transformations dwindled across the curriculum and demonstration gradually disappeared from the middle school syllabus. After a brief reappearance in the 10th grade in the 2000s, the congruence and similarity of triangles returned to the middle school curriculum in 2016. However, as the teacher training sessions led by the group Géométrie de l'IREM de Paris (in which Marie-Jeanne Perrin-Glorian participated) on this theme have shown, teachers hardly use congruent triangles as demonstration tools: most of them never encountered them when they were pupils or student teachers and the first generation of textbooks do not use them as demonstration tools. Moreover, it seems that currently some theorems are proved in the courses but without any concern for overall coherence. Therefore, as we have seen during the teacher training sessions that we organized, many middle schools study similar triangles before the proportional segments theorem that would help to prove the relationships between equality of angles and proportionality of sides of similar triangles.

For the last four years, the same group has been working on the construction of a progression for geometry at the middle school level, based on the congruence of triangles with, underlying Euclid axiomatics that have been adapted a little (Groupe Géométrie de l'IREM de Paris 2021). It is a continuation of the work done in the research group on geometry at the IUFM Nord-Pas-de-Calais (henceforth called the Lille Group) in which we both worked in the 2000s and which led notably to a book that was published in 2020 (Mathé et al. 2020). Indeed, Euclid's axioms and first theorems model graphical objects and actions with the instruments whose use we codify in the geometry of tracing.

1.3. Geometry of tracing, a possible link between material geometry and Euclid's geometry?

The word "tracing" evokes instruments and the material figure, but geometry of tracing as we understand it is different from material geometry. Indeed, it corresponds to the search for a possible link in learning between tracing lines with instruments and the notions of abstract geometry. It reflects the fact that the objects and axioms of Euclid's geometry are not chosen at random, and that they can be seen in large part as a theorization of figure construction, in particular figures that can be constructed with a ruler and a compass (by tracing lines and circles). Indeed, the first propositions in Euclid's *Elements*[2] concern the construction of the equilateral triangle and the transfer of length. The criteria for the congruence of triangles follow and validate many constructions. For example, of the 48 propositions in Book I, 14 concern constructions. In this section, we will specify some essential points for the geometry of tracing and show how it can prepare for the transition into theoretical geometry.

1.3.1. *Figure visualization and figure restoration*

More than 20 years ago, our Lille Group tackled the question of finding a coherent progression of geometry throughout compulsory education, from kindergarten to the end of middle school, mainly in Cycle 3 (9–12 years old)[3], which, now including the 6th grade, must manage the start of the transition from the geometry of tracing with instruments to the deductive geometry. At the beginning, we emphasized the way to look at the figures: from the draftsman's view to the analytical view that the geometric visualization of the figures requires. Based on Duval's work, we distinguished and operationalized different visualizations of geometric drawings (Perrin-Glorian and Godin 2018; Mathé et al. 2020): "surface, contour, line, point" visualizations[4].

In order to work on the geometric view of figures, that is, to see relationships between smaller elements (lines, points, circles) constituting the figure, we have identified a fundamental situation and its didactic variables: figure restoration. The action situation consists of reproducing a model figure from a starting figure (part of the figure already restored), with identified instruments, to which a cost can be attributed in order to develop the procedures. Many examples are presented in previous papers (including Keskessa et al. 2007; Perrin-Glorian and Godin 2014; Mathé et al. 2020). We will see more in the following. Note that the reproduction of figures corresponds to the case where there is no starting figure.

---

2 *The Elements of Euclid* (Vitrac translation) (1990). PUF.
3 In France, compulsory schooling is organized in cycles: cycle 1 (3–6 years); cycle 2 (6–9 years); cycle 3 (9–12 years); and cycle 4 (12–15 years). Primary school is from 3 to 11 years old and middle school from 11 to 15 years old (thus it includes the last year of cycle 3 and the whole of cycle 4).
4 We refer the reader, for example, to the text by Christine Mangiante in this same book, for a more detailed illustration of these notions.

1.3.2. *The geometrical use of tracing instruments, a first step to make sense to an axiomatic*

The definition of "geometry of tracing" aims to identify a field of practices and discourses that can be held on physical geometric figures to facilitate the entry into theoretical geometry. The objects are those of GI: physical geometric figures and the instruments that allow them to be produced. However, the practices are not quite those of GI. There are differences with GI at the level of the ostensives, since measuring instruments are excluded, and at the level of the non-ostensives which allow for the use of the ostensives to be regulated[5], in particular the rules for the use of instruments which we present in the following. The geometry of tracing, as we define it, is built around the reproduction of figures. The instruments available are an important variable in these situations. We call instruments *physical artifacts* that allow us to reproduce the figure, excluding measurements (we transfer quantities without measuring them). In Euclid's geometry, we have ratios for quantities but not numbers. Measurement issues are also important in relation to numbers, but we do not include them in the geometry of tracing. As we are also interested in cycle 2 (6–9 years), or even in kindergarten (3–6 years), we consider among the tracing instruments not only the usual instruments, non-graduated ruler, set square, compass, but also various materials (templates or stencils of plastic, wooden or cardboard shapes, etc.) which refer to visualizations of the figure as assemblies of surfaces or tracing around the contour of the surface.

We have analyzed the operations of figure restoration and construction, and identified theoretical instruments (without the limitations of material instruments) that each represent one of the elementary functions that allow for the reproduction of reproducible figures with a ruler and compass: a ruler[6] (not graduated) to draw straight lines, as long as we want; length conveyor; length bisector that allows us to take half a length; angle conveyor[7], for example, in the form of a template with the special case of the set square as a right angle conveyor; compass to draw circles. Except for the first tracing where there is no starting figure, the geometrical use of these instruments is constrained by rules that can be seen as a kind of practical

---

5 For further reading on ostensives, see Bosch and Chevallard (1999).

6 In order not to lengthen the text, and as is customary when speaking of the geometry of the ruler and the compass, we call the ruler the straight-line tracer and the compass the circle tracer. Each of the theoretical instruments can be implemented in different ways; the material instruments have limitations, for example, the length of the ruler and the spacing of the compass. The theoretical instruments evolve as the pupils' knowledge develops.

7 The angle conveyor can also be materialized by a circle on transparent paper with a half-line starting from the center.

axiomatics that could be considered as the beginning of a concrete version of Euclid's axiomatics, with the addition of the fact that lines have no thickness and that points are obtained by the intersection of lines:

– To set up the ruler, you need two points or a segment already drawn.

– To set up the length conveyor, you need a line already drawn and a point on this line; the length is transferred to the line on one side or the other of this point.

– To take the middle of a segment, we transfer the length of this segment on the length bisector, and then we transfer the half-length onto the segment from one end.

– To place the angle conveyor, you need a straight line and a point on this line (or a half-line): the top of the template (or the center of the circle) is placed on the point and one side of the template is placed on the line; the angle is plotted on one side or the other (four positions if you have a straight line, two positions if you have a half-line).

– To draw a circle with the compass, you need a point upon which to place the point of the compass (the center of the circle) and a point along the circle, or a length that gives the distance of the compass.

However, there is still an axiom missing to draw the parallel or the perpendicular to a line passing through a point outside the line: you must be able to transfer the angle without knowing the vertex, by sliding a side of the template on a line. This is a kind of axiom for parallels:

– While dragging one side of a corner template onto a straight line, we can draw parallel lines with the other side of the template.

With these rules for instrument usage, we can build all the figures of plane geometry encountered in elementary and middle schooling. These functions can likewise be found in dynamic geometry, software such as Cabri Geometer® or GeoGebra®.

As the pupils' knowledge evolves, the number of instruments can be reduced: the compass can quickly replace the length conveyor, then the segment bisector, then the square and the angle conveyor. The teaching of geometry from 6 to 12 years of age must, in our opinion, aim at a first conceptualization of objects, relations and geometrical properties based on these rules of instrument uses. We believe that the explication of these rules for constructing physical geometric figures helps in the conceptualization of abstract geometric objects found in theoretical geometry.

1.3.3. *Distinguishing between the hypothesis and the conclusion*

Pluvinage and Rauscher (1986) already showed how the construction of geometric figures leads to deduction. We want to show how figure restoration contributes to this insofar as it leads to asking the fundamental questions for problem solving: "What do we already have? What are we missing?", and similar to those, that which must be asked in a demonstration: "What do we know? What are we looking for?" Indeed, in a figure restoration, there is a starting figure that we have to identify on the model, which can be put in parallel with the identification of data in a demonstration. We seek how to trace a difference figure, the one that will supplement the starting figure to obtain the model figure. In order to draw it, we must connect the lines we are looking for to those we already have through an acceptable construction process, that is, a geometrical use of the instruments at our disposal. In the same way, in a demonstration, we try to link the results we want to prove to the data of the problem (the hypotheses) by using results already validated (theorems, definitions or results already demonstrated).

Let us follow an example. We want to replicate the following figure (Figure 1.3(a)) using the trapezoidal starting figure provided at a different size and in a different orientation (Figure 1.3(b)), on the same sheet. We are missing one arc and one segment.

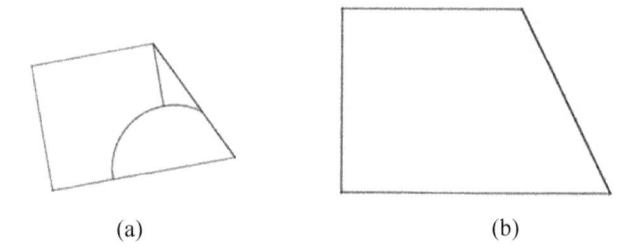

(a)   (b)

Figure 1.3. *A figure (a) to be replicated and (b) its starting figure*

The missing end of the segment is on the arc. To replicate the arc, you need its center and radius or its center and one point. To replicate the segment, you need the other end. We need to connect the elements we are looking for to the starting figure. To do this, we must make assumptions about the model and check them with the instruments. To find the direction of the segment that connects a vertex of the trapezoid to the arc of the circle, we can check on the model that it is perpendicular to the small base of the trapezoid, but to find the center of the circle, we must see the point where the segment meets the arc as a point of intersection, and to do this, we must extend the segment on the model to the large base of the trapezoid in order to create a square (Figure 1.4(a)). A square appears – this is established by checking the equality of the sides and the right angles. We thus find both the direction of the missing segment and the center of the circle which is the point created (the fourth vertex of the square). It remains to see that the circle passes through the middle of the side of the square or that its radius is half the side of the square which can be done with an internal length transfer[8] plotted onto the model.

Extending the arc to obtain a semicircle, one end of which is the vertex of the acute angle of the trapezoid (Figure 1.4(b)), will allow the circle to be drawn without the need to find a midpoint. Indeed, the radius is also the short side of the right angle of the right triangle drawn at the same time as the square, which is one-third of the large base of the trapezoid.

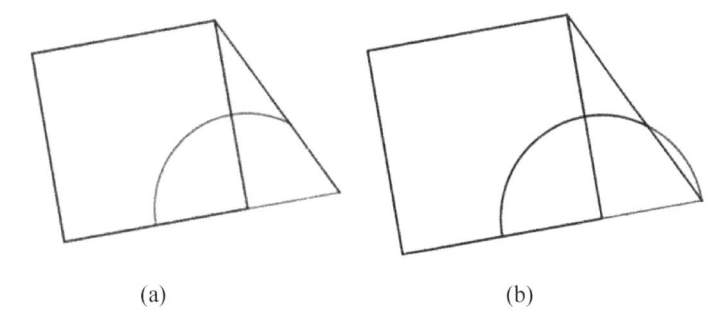

(a) (b)

Figure 1.4. *Adding traces on the model figure*

To draw the fourth vertex of the square on the starting figure, two procedures are possible: either the perpendicular to the large base of the trapezoid passing through the appropriate vertex of the trapezoid or a transfer of the length of the side of the square to the large base of the trapezoid (Figure 1.5). We can then draw a semicircle, with this point at its center, passing through the vertex of the trapezoid. Asking for a restoration at a different scale forces us to see the relationships between the lines we trace instead of just reproducing lines.

---

8 In figure restoration, we distinguish between length transfers within a figure (e.g. to verify a midpoint or a ratio of lengths) which correspond to a search for properties of the model that will have to be restored, and transfers from the model to the figure which correspond to information on the length itself.

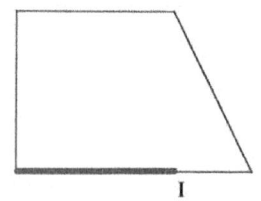

Figure 1.5. *Tracing on the starting figure*

1.3.4. *Restoration, description, construction of figures and geometric language*

Figure restoration is a form of figure reproduction: we must produce a figure identical to the figure given as a model, in a sense that needs to be specified (must the figure be superimposable on the model or are certain differences to be specified accepted?). We must look for properties on the model that must be reproduced. The reproduction of a figure by someone who has both the starting figure and the model does not require the use of language, unlike figure constructions or descriptions. Describing a figure requires producing sufficient information from the figure to characterize it unambiguously. By description we often mean a written text, but we can also imagine an oral message or even graphical language. A construction program is a particular description in the form of a text that gives a procedure for constructing the figure with instruments. Conversely, the construction of a figure requires producing a figure from a description of this figure (most often a text).

Reproducing a figure is generally a matter of material geometry: the object of work is the physical geometric figure and the validation is done by perceptive comparison of the model figure with the produced figure, using tracing paper or instruments. The description and construction of figures also concerns physical geometric figures, but the didactic contract can be based on material geometry or theoretical geometry. Therefore, the description of figures with a view to their construction and the validation processes that it generates seems to be a transition mechanism between material geometry and theoretical geometry.

The question that pupils ask themselves when restoring a figure is: "How do I construct the (physical) figure with instruments?" We believe that it is possible, through the language work of description and construction, based on the reproduction of a figure, to guide pupils towards two new kinds of questioning. The first concerns the designation and definition of these objects: "What information is necessary and sufficient to characterize a geometric object or a given figure?" More than questions of vocabulary, we will see that the designation of geometric objects is intimately linked to their definition. This questioning constitutes a learning issue in geometry at school and in middle school, the importance of which we believe is underestimated. The second line of questioning is directed towards introducing pupils into the activity of proof. It is a matter of engaging them in the following question: "How can we, given what we know (notably through a construction program), establish other properties using what we have?"

Questioning the role of language in the processes of learning and teaching geometry is, of course, not new. This was studied in the foundational work in mathematics education about geometry (Laborde 1990; Berthelot and Salin 1992; Fregona 1995). More recently, it has notably been at the heart of the work that Anne-Cécile Mathé developed with Caroline Bulf and Joris Mithalal (Bulf et al. 2014), of Edith Petitfour's dissertation, directed by Corine Castela and Marie-Jeanne Perrin-Glorian (Petitfour 2015, 2017a), and, for example, the current work of Joris Mithalal and Marianne Moulin (2019) on construction programs. Each approaches the question from a slightly different angle that nonetheless has strong complementarity.

For some years, the research, within two of the IREM groups of Clermont Ferrand, on situations which confront pupils with the questions mentioned above and make their exploration necessary, has led Anne-Cécile Mathé to put the situations of reproduction of figures back into the framework of the dialectics of action, formulation and validation (Brousseau 1997b). If Berthelot and Salin (1992) or Fregona (1995) already explored the potentialities of these situations, we study them today with a slightly new outlook, enlightened by the work of the Lille Group and the perspective of developing the transition between material geometry and theoretical geometry. We develop and illustrate this point now, trying to show how the dialectics of action, formulation and validation fit into our approach.

1.4. Dialectics of action, formulation and validation with regards to the reproduction of figures with instruments

Restoration, as well as the reproduction of figures, is an action situation. To solve such a problem, it is necessary to mobilize situational knowledge, in the sense of knowledge in situation. In order to access the practice of geometry, action is not enough; pupils must also feel the need for an appropriate and shared language, as well as the need for intellectual proof to validate their assertions.

1.4.1. *Formulation situations and possible variations*

In the formulation situations, the principle is to prevent direct action, by placing the model figure and the support of its restoration at a distance. These situations are designed to make it necessary to formulate the situational knowledge involved in the reproduction of figures, which will then become a matter of recognizing, identifying, breaking down and reconstructing it through a linguistic system (Brousseau 1997a, p. 7). Formulation is necessary if direct action on the environment – in this case, taking information about the model figure – is prevented; for example, because the model figure is far away, in space or in time, from the place or time where it is to be restored. If the sender and receiver of the message are different people (formulation for others), the two players must cooperate. The formulation must hence be more explicit and shared. Another key variable in these situations is the means used to communicate. This may be graphic language (e.g. a freehand drawing or a coded figure highlighting the information needed for restoration) or verbal language. When verbal language is used, it may be oral or written. Finally, we can envisage making choices with regards to the constraints imposed on the nature of the verbal discourse produced, for example, between technical language and geometric language.

In addition to the choice of the model figure and the available instruments, we can envisage a wide variety of formulation situations around the reproduction of figures. Further to this, we could likewise take into account other didactic variables in the development of these situations, considering gestures as a form of language, for example. The ones we have just mentioned already offer a wide range of situations, which we can represent as follows, shown in Figure 1.6.

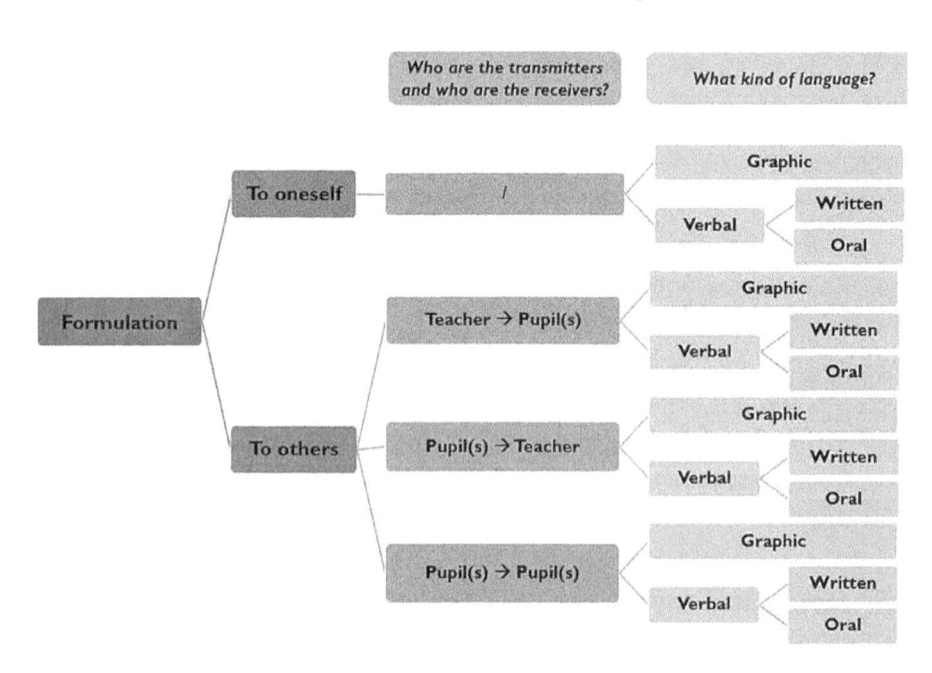

Figure 1.6. *Variation of formulation situations*

Each choice of variable brings into play different types of situational knowledge. We cannot carry out a detailed a priori analysis for each case; however, just considering this great variety of formulation situations hints at the games that can be played using the variables of these situations, in turn encouraging progressive pupil learning based on the construction of a geometrical language in order to carry out an analysis and characterization of the geometrical objects and the relations involved in the reproduction of figures. It is along this line of thinking that we focus upon within our IREM groups. The article by Mathé et al. (2021) provides an informative overview on the designation of the circle.

1.4.2. *Validation situations*

In action situations as well as in formulation situations, the evaluation of the tracing strategies, or of the formulations aiming at communicating these strategies, comes under a pragmatic proof: the superposition between two physical geometric figures, one of which is built by the pupil, directly or after the exchange of a message. The validation situations are part of an intellectual proof perspective.

The objective is to work on the construction of explanations, the identification of reasons for which the strategy used and formulated effectively allows, for certain, the reproduction of the figure. These situations are based, in particular, on written verbal formulations. The questions posed are: "Does the text give the necessary and sufficient information to restore the figure? Does it allow for the construction of a physical geometric figure that does not verify the geometric properties of the model figure?" The object of the work is thus here the text, of which the physical geometric figure becomes an auxiliary representative. Proof procedures are undertaken, in which the physical geometric figures produced can be used as illustrations or counterexamples to invalidate a message.

In the following section, we will present two examples of progressions articulating figure restoration situations and formulation situations carried out in the end of elementary school and the beginning of secondary school, aiming at a first conceptualization and then the characterization of fundamental notions of geometry. In the final part, we will discuss possible validation situations and their potentialities in terms of entry into proof activities in middle school.

1.5. From tracing to the characterization of objects and geometric relationships

1.5.1. *On the concepts of segments, lines and points*

We present here some elements of a progression implemented in Julien Ribennes's 5th grade class (10–11 years) in Chamalières (Puy-de-Dôme)[9].

1.5.1.1. *Targeted knowledge and skills*

The notions of segments, lines and points appear today in Grade 1–3 (6–9 years old) school programs (BOEN no. 31 of July 30, 2020). However, if the programs evoke the idea of using geometric objects and properties as tools for solving problems involving physical geometric figures, the emphasis is still mainly on vocabulary and notation. The work of the Lille group showed the difficulty of the encounter with the conceptual objects of lines and points, which supposes a complex and specific visual analysis of the figures in terms of the visual articulation of surfaces, contours, lines and points. The sequence of introducing the notions that we have set up in 5th grade aims at the following learning objectives:

– Using the relationships of alignment and incidence between segments, lines and points in the reproduction of figures, that is: be able to analyze a composite figure; perceive the networks of lines supporting given figures; see vertices as points of intersection for lines, and identify the alignment properties of points and/or segments, properties of points belonging of lines and/or segments, inclusions of segments in lines and/or segments; be able to mobilize these properties to restore a composite figure with the help of a non-graduated ruler.

– To put into words an initial conceptualization on the notions of lines, segments, points and the relations of alignment and membership, based on a geometrical use of the ruler. A segment is represented by a straight line (that can be drawn with a ruler) that has two ends; a straight line is represented by a straight line that can always be extended; when two straight lines intersect, they form a point; three points (a point and a segment, or two segments, respectively) are aligned when they are on the same line.

– Be able to designate segments, lines and points and describe the properties of alignment and belonging that are used. In particular, this involves questioning the prime objects needed to characterize and designate a line or a segment. At the end of elementary school, in accordance with the programs, we address the question of notations that designate without ambiguity segments, lines and points.

---

9 These situations are the fruit of a collective reflection, carried out within the group "Teaching geometry at school" of the IREM of Clermont Ferrand, led by Anne-Cécile Mathé.

1.5.1.2. *Elements of progress*

The pupils are first confronted with situations of figure restoration with cost on the instruments making necessary an articulation between visualizations of surfaces, contours, segments, lines and points of the composite figures, and the mobilization of alignment and belonging properties. The sharing and conclusion phases around these situations are the place for verbalizing the analysis of the figure and the instrument procedures. This work allows the teacher to set up an initial geometric language (segment, line, point, extend, trace, etc.) and to gradually build up an initial institutionalization of geometric knowledge concerning the relationships of alignment and incidence, based on the geometric use of the non-graduated ruler. The next part of the proposed progression extends this work. It is thus a question of problematizing the designation and characterization of these objects and relations. We begin by confronting pupils with oral formulation situations, playing on the distribution of the roles of sender and receiver (pupils to teacher, teacher to pupils, pupils to pupils). These first situations are a place for appropriating a foundational language, making it possible to describe segments, straight lines, points and relationships of alignment, belonging and incidence.

The question of how to name these geometric objects is discussed and the appropriate notation conventions are introduced. The pupils are then offered written verbal formulation situations, of which the following is an example (Figure 1.7). The sender receives two sheets: the first contains the model and the starting figure; the second contains the starting figure only and leaves room to write a message for the receiver who will have to restore the figure according to these indications.

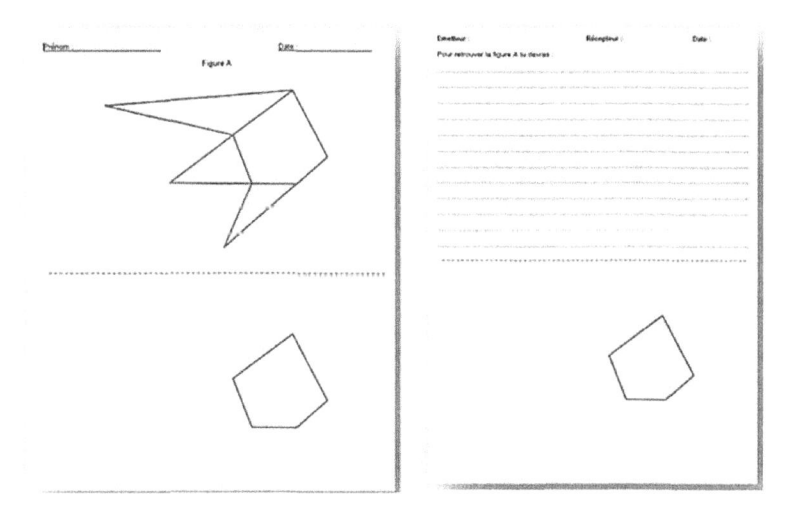

Figure 1.7. *Example of support for a situation where the formulation to others is written*

Figure 1.8 shows an example of a message produced by a pupil and the realization by the receiver along with their comments. The figures are given without notation. It is the senders who name the starting figure points on the message they are transmitting.

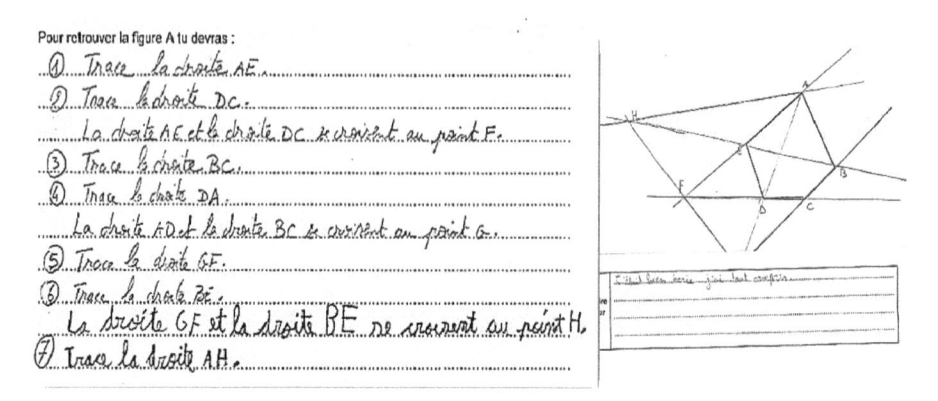

Figure 1.8. *Example of pupil productions*

This valid production makes it possible to measure the considerable jump that represents the passage from a description of instrument procedures for reproducing figures, even in geometric language, to the production of a text that is necessary (and sufficient) to characterize a figure by stating geometric relations between lines and points (subunits of dimensions 1 and 0). By necessity of the situation, the pupils are confronted with the designation of figurative units and the formulation of relations between these figurative units. How to designate without ambiguity a segment, a line and a point, in order to communicate around these objects?

These situations are the place where the first notations are introduced, which appear as tools and are related to questions of characterization: two points are necessary to draw a line with a ruler, so two points are enough to unambiguously designate and characterize a line. The question of the order of appearance of the figurative units is also raised: which objects must be available to construct others? For example, to construct the line (GF), we must first construct the points G and F.

As we can see here, the naming of geometric objects goes beyond the mastery of a vocabulary. It involves an ability to characterize these objects. The formulation situations seem to be extremely rich since they lead to exploring the question of what is the necessary and sufficient information needed to characterize a physical figure through geometric objects, properties and relations. They thus encourage the progressive engagement of pupils to practice geometric reasoning.

Activities based on construction programs are common in this grade in France, although they most often consist of proposing that pupils construct a figure from a given program. We think that such an activity does not confront the pupil with the necessities and constraints that weigh on geometric language. Thinking of construction programs as the product of formulation situations also makes it possible to envisage a learning progression.

Therefore, the prior confrontation of the pupils with oral verbal formulation situations seems to facilitate the entry into the writing of construction programs by circumventing the pupils' difficulty of expression through writing. Most importantly, whereas written verbal-other formulation situations require producing a complete text on the first try, anticipating the constraints on the text, oral interactions allow for direct feedback from the receiver as instructions are given. Through her work, Petitfour (2017b) demonstrates the potentialities of this type of situation.

1.5.2. On the notion of perpendicular lines

The following example was implemented in Claire Rosalba's 6th grade class in Pont du Château to work on perpendicularity.[10]

1.5.2.1. Targeted knowledge and skills

The 2020 programs (BOEN no. 31 of July 30, 2020, p. 97) stipulate that 5th grade pupils must be able to "recognize and use a few geometric relationships" including "perpendicularity". More specifically, the *Repères annuels de progression* (2019, p. 7) propose that, starting in 5th grade, pupils "draw with the set square the perpendicular line to a given line passing through a given point that may be outside the line". It also states that "specific vocabulary is used from the beginning of the 3rd grade to designate objects, relationships, and properties" (p. 8, *translated by author*).

As with the previous theme, our learning objectives articulate three dimensions of geometric activity: a conceptualization of the notion of perpendicular lines, a geometric use of the set square, the designation and description of the perpendicularity of lines. The aim is to encourage the pupil ability to:

– analyze a physical geometric figure geometrically, perceiving in particular right angles and perpendicular relationships between segments;

---

10 This work is the fruit of a reflection carried out on "Problem solving in middle school" by the IREM group of Clermont Ferrand, led by Aurélie Roux.

– restore a physical geometric figure by drawing segments or lines from two points or a segment, by constructing perpendicular lines to a given line, passing through a given point;

– describe two perpendicular lines and be able to use geometric language, such as "the line perpendicular to the line d, passing through point A" in a situation involving the ability to characterize perpendicularity as a ternary relationship, between two lines and a point.

1.5.2.2. *Initial restorations of figures involving perpendicularity*

The pupils have already encountered figure restorations during previous work on the notions of segments, lines and points. They have begun to progressively develop an ability to analyze composite figures, and to extend segments into straight lines.

The didactic challenge of the action situation now consists of having the pupils geometrically analyze the material figure, by perceiving not only the alignments and intersections but also the right angles (there are two in the example) and the perpendicular relations between segments. They will then have to restore the figure by drawing segments or lines that verify the properties identified on the model, in particular by constructing lines perpendicular to a given line, passing through a given point. Figure 1.9 shows an example of a model figure and a starting figure, with the only instruments available being a set square and a non-graduated ruler. On the left is the material provided, on the right is the enrichment of the model and the construction on the starting figure.

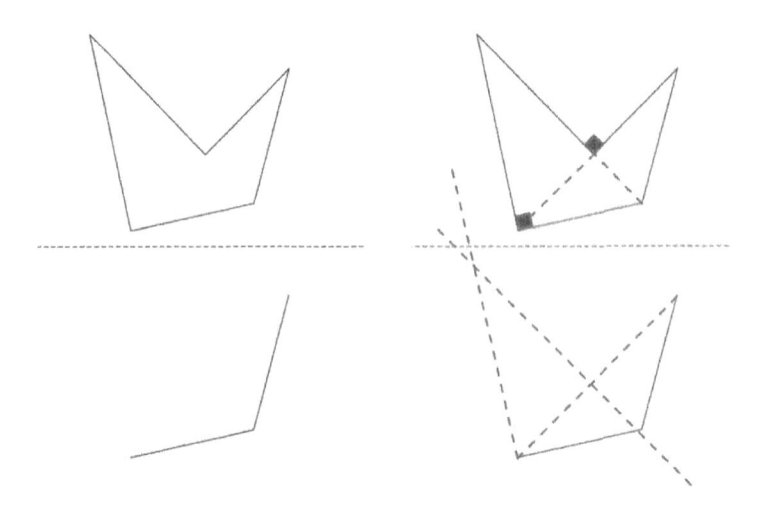

Figure 1.9. *Example of figure restoration involving perpendicularity*

Drawing segments and lines perpendicular to a specific segment or line through a specified point is physically accomplished by placing one's ruler on the specified line or segment and then dragging one of the sides adjacent to the right angle of the set square along the line (the segment) until the specified point coincides with the other side adjacent to the right angle (Figure 1.10).

The collective sharing and conclusion phases are here as well as the place for verbalizing the restoration strategies and installing a preliminary geometric language that makes it possible to describe the geometric objects and relations mobilized, based on the instrumented procedures: use of the words "perpendicular lines" or expressions such as "We have drawn a line perpendicular to … passing through …". However, the rest of the sequence shows us the extent to which mobilizing this type of language for a given situation, that is moving from a description to a characterization of the relationship, represents a considerable cognitive leap for pupils. Let us take a look at what happens when we ask pupils to communicate a written verbal formulation to another person.

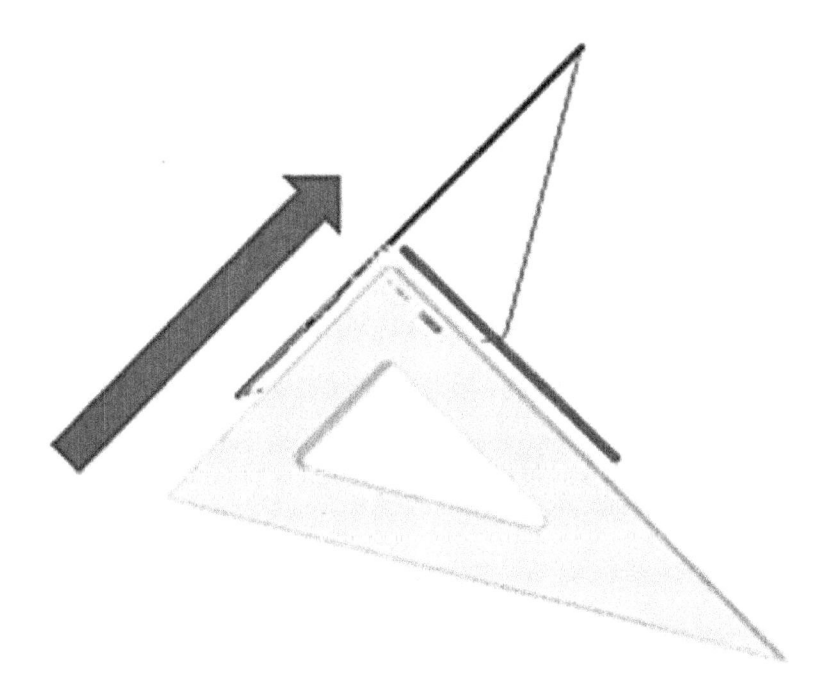

Figure 1.10. *Drawing a line or a segment perpendicular to a specified line*

1.5.2.3. *Communicating a written verbal formulation to others situation*

The task now is to restore a figure through a communication situation between sender and receiver. The sender receives the model figure and the starting figure on the same sheet of paper (Figure 1.11) and must write a message to the receiver. The receiver has to restore the model figure from the starting figure and this message. Both the sender and the receiver have a non-graduated ruler and a set square.

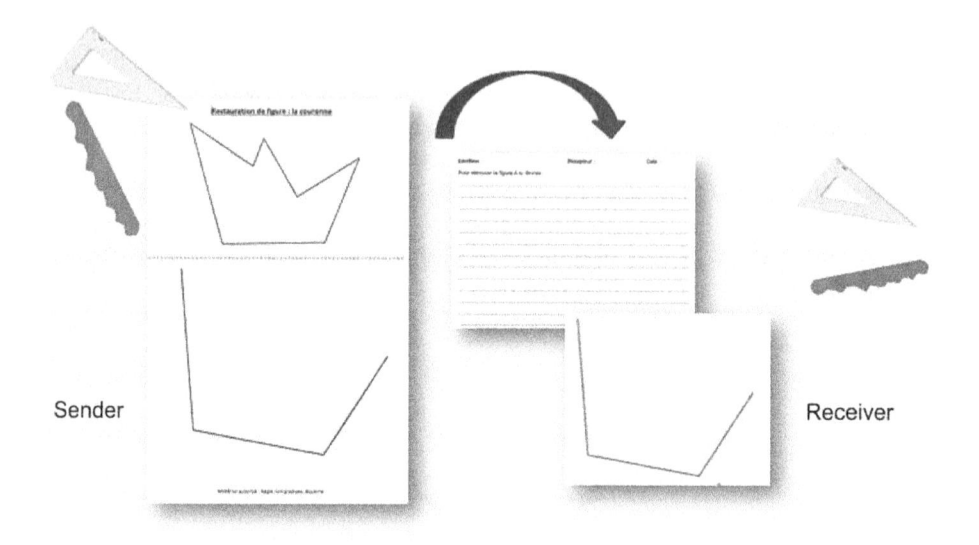

Figure 1.11. *A situation of verbal or written formulation communicated to others*

Writing a message requires first restoring the figure oneself, making visible a network of segments underlying the construction of the figure, alignment relationships between points or segments and points, and identifying a right angle (Figure 1.12). The analysis of the figure must then be prioritized, and the geometric objects and relationships used must likewise be identified.

Restoration of the figure: the crown

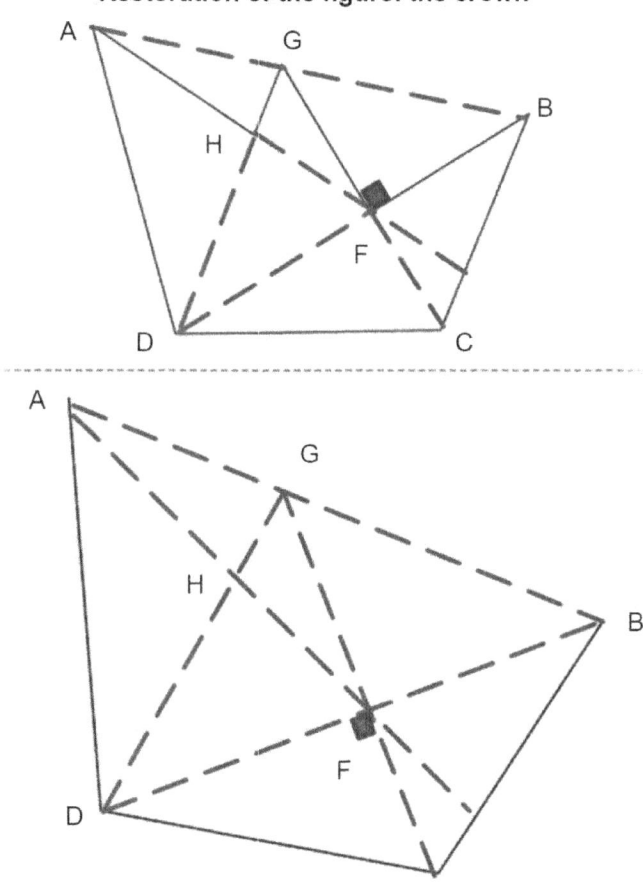

Authorized material: non-graduated ruler, square

Figure 1.12. *Reproduction of the figure (model and starting figure completed)*

Producing such a text is extremely complicated if we do not decide to name the points. We therefore introduce, as pupils would, the names of the points, taking care to transfer the same names to the starting figure as on the model. Since the points A, B, C, D and the segments [AD], [DC] and [BC] are given by the starting figure, a valid communication could be the following:

– "Draw segment [AB]."

– "Draw segment [DB]."

– "Construct the line perpendicular to segment [DB] through C. This line intersects segment [DB] at F and segment [AB] at G."

– "Draw segment [GD]."

– "Draw segment [AF]. This segment intersects segment [GD] at H."

At this stage of the progression, pupils are now trained to analyze composite figures and to perceive the underlying networks of lines and segments and the properties of alignment and perpendicularity that link them. A majority of the pupils manage to restore the figure with a non-graduated ruler and a set square. However, in this first situation of formulation, many pupils have great difficulty in putting into words a designation for the perpendicular line that allows it to be characterized, and thus to be drawn by a classmate. The example of pupil explanation (Figure 1.13) shows the difficulty in defining (CG) from (DB).

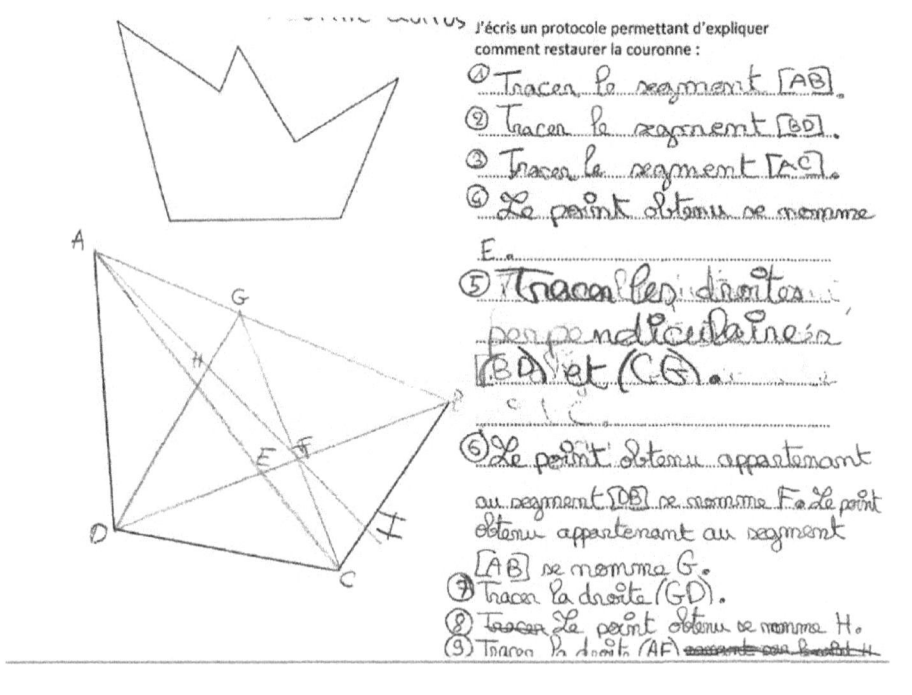

Figure 1.13. *An example of a pupil's explanation*

However, testing the messages produced allows pupils to become aware of their shortcomings. During a validation phase, language work can then begin on the characterization of the perpendicular line in question, by interacting with their drawing using a ruler and a set square.

As the pupils and the teacher interact through language, thanks to the feedback provided by the situation, the pupils gradually become aware that characterizing the line is based on a ternary relationship between two lines and a point. It then becomes clear that there is an infinite number (or at least a multiplicity) of lines perpendicular to a given line. The expression "the line d perpendicular to the line d' passing through point A" is then instituted as a tool to solve the problem.

This work can give rise to a written trace of the type:

> There are infinitely many lines perpendicular to the line (BD). There is only one line perpendicular to the line (BD) and passing through the point C. We can say "The lines (BD) and (CG) are perpendicular at F" or "The line (CG) is the line perpendicular to the line (BD) passing through C". We then can write "(BD) $\perp$ (CG)".

1.6. Towards proof and validation situations in relation to figure restoration

Recall that our project is to explore ways through which to bring pupils to encounter two types of questions (section 1.3.4). The first type, mentioned in section 1.5, concerns the necessary and sufficient information required to characterize a given geometric object or figure, in connection with naming questions. The second type of question concerns directly introducing the proof activity. In this section, we propose some lines of thought, based on examples of validation situations that have yet to be tested.

1.6.1. *Equivalence between two construction programs and the need for proof*

Let us go back to the figure restoration proposed above (Figure 1.3). There are several ways of restoring the figure. Each of these instrumented procedures could, in a written formulation situation, give rise to a different type of message. Here are two examples, only varying according to the procedure for constructing the square.

Message 1 (corresponding to the restoration shown in Figure 1.14, the smaller figure represents the properties identified on the model by the sender, the larger the properties used by the receiver to complete the starting figure according to the message):

- "Draw a line perpendicular to the line (AB) passing through B."
- "It intersects the segment [DC] at E."
- "Draw the circle with center E passing through C."
- "It intersects segment [BE] at J."

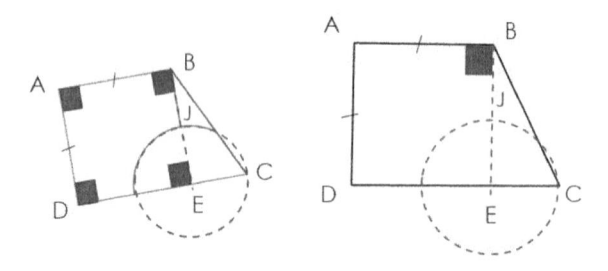

Figure 1.14. *An example of restoration*

Message 2 (corresponding to the restoration shown in Figure 1.15):
- "On the segment [DC] place a point E, such that DE = AD."
- "Draw the circle with center E passing through C."
- "The line (BE) intersects the circle at a point J. Draw the segment [BJ]."

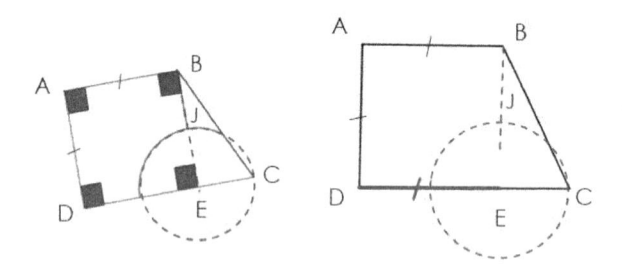

Figure 1.15. *A second example of restoration*

The plurality of operative messages can be very rich in the classroom. It naturally leads to questions such as "Why are both construction methods equivalent?" In a first visual analysis of the figure, we immediately identify a square ABED. If we rely on these construction programs, why are we sure that we end up with a square[11] in both cases? The procedures used can thus lead to isolate the construction of the square for which we have encountered a need for proof.

1.6.2. *Validation situations involving programs for the construction of a square and introducing a proof process*

Consider the construction of the square from two consecutive sides.

1.6.2.1. *Proving the invalidity of a construction program, the material figure as a counterexample*

If we have two sides as a starting figure, we have three vertices of the square. To find the fourth, many pupils place the set square as shown in Figure 1.16, in order to perceptually complete the square. To invalidate this procedure, we can propose the situation in Box 1.1.

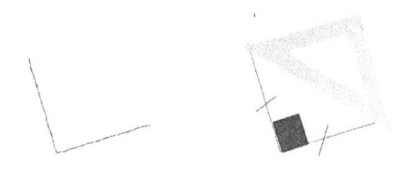

Figure 1.16. *An invalid use of the set square*

> A classmate produced the following message to construct a square from the given starting figure: "To draw both sides of the square, draw a right angle at the missing vertex." Does the message construct a square?

Box 1.1. *A situation to propose*

To invalidate the message, we wait for the production of counterexamples like this one (Figure 1.17). The drawing is a mathematical proof that shows that two opposite angles of a quadrilateral can be right without the other two angles being right. We can move the set square around the vertices (be sure to note that the sides will not be equal either).

---

11 We assume the square is defined as a quadrilateral with four sides of equal length and four right angles.

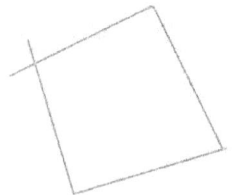

Figure 1.17. *A possible counterexample*

The work here focuses on the text that defines the figure. The drawing becomes a secondary representation that completes the text. Here, several types of drawings are suitable: representations of squares but also of other quadrilaterals that have only two opposite right angles. The production of the counterexample is an opportunity to make explicit an element of the practical axiomatics that allows the message to be invalidated: "To place the set square, you need a supporting line to place one side of the right angle and a point to place the vertex of the right angle, or the second side of the right angle."

1.6.2.2. *Proof for the validity of a construction program*

Based on this practical axiomatic, we can produce Message 2: "Trace the perpendicular lines to each of the given sides, each passing through one end of the side." We may then note that the starting figure and the construction program fix only three right angles (one is given by the starting figure; the other two are constructed) and two equal consecutive sides.

Therefore, pupils can be led to explore the following question: "Why are we sure we are getting a square?" The production of physical geometric figures can help to convince the validity of the message (Figure 1.18).

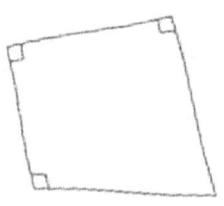

Figure 1.18. *A freehand-drawn figure*

However, this time an example is no longer enough. How can we be sure that it will always work? Can we find reasons why the fourth angle is a right angle and that the sides are equal pairs? We will then have to resort to statements and theorems, which at this level, will be admitted but that we can reuse to prove other statements:

> If two lines are perpendicular to the same line, they are parallel to each other.

> If two lines are parallel and a third line is perpendicular to one of them, it is also perpendicular to the other.

1.7. Conclusion

The rules for the geometric use of instruments aims to set up a technical geometric language, in the sense employed by Petitfour (2015, 2017a), and to highlight relations between material traces that correspond to relevant relations between the geometric objects of which they are the representations in the theoretical geometric space. The formulation situations extend the figure restoration situations through work focusing more on the language dimension of the geometric activity. As we can see from the two examples, beyond the issues related to the mastery of a vocabulary, the naming of geometric objects brings into play an ability to characterize these objects. We believe that such work can provide the setting for a progressive transition between material geometry and theoretical geometry. It engages pupils in geometric reasoning on texts and questions on the information necessary to characterize a given geometric object or figure.

The extent of the difficulties encountered by the pupils in this type of task, far from discouraging us, is a sign of the complexity of the naming and characterization activities undertaken. It seems important to us to take hold of these questions as it is necessary to think about this learning over a long period of time. Current work shows the ability of pupils to move from technical geometric language to geometric language and from material objects to the ideal objects of the model through well thought-out and well-managed formulation situations. Moreover, exploring the question of equivalence between construction programs, finding reasons for the invalidity or validity of messages seems to provide the opportunity for rich activities, which, even beyond 5th grade, could allow us to accompany pupils in a modification of the object of study, and to engage them in proof checking procedures. These reflections are at the center of our current interests. The experiments and their analyses remain to be continued.

The use of such situations in ordinary teaching presupposes the training of primary school teachers, in particular through in-service training, but also and above all in secondary school: the geometry training of primary school teachers relies heavily on the relationship to geometry that they themselves experienced in secondary school. Sixth grade, the pivotal point between cycle 3 (9-12 years old) and cycle 4 (12–15 years old), seems to be an essential phase in which to study the possibilities of implementing formulation and validation situations, for the restoration and construction of figures on a larger scale, allowing for a more efficient introduction of pupils to proofing in geometry.

1.8. References

Berthelot, R. and Salin, M.-H. (1992). L'enseignement de l'espace et de la géométrie dans la scolarité obligatoire. Thesis, Université Bordeaux 1.

Bosch, M. and Chevallard, Y. (1999). La sensibilité de l'activité mathématique aux ostensifs. Objet d'étude et problématique. *Recherches en didactique des mathématiques*, 19(1), 77–124.

Brousseau, G. (1997a). La théorie des situations didactiques. Cours donné par Guy Brousseau à l'occasion de l'attribution à celui-ci du titre de docteur Honoris Causa de l'Université de Montréal [Online]. Available at: http://guy-brousseau.com/1694/la-theorie-des-situations-didactiques-lecours-de-montreal-1997/.

Brousseau, G. (1997b). *Theory of Didactical Situations in Mathematics. Didactique des mathématiques 1970-1990*. Kluwer Academic Publishers, Dordrecht.

Brousseau, G. (2000). Les propriétés didactiques de la géométrie élémentaire. *Actes du séminaire de didactique des mathématiques*, Rethymon, Université de Crète [Online]. Available at: https://hal.archives-ouvertes.fr/hal-00515110/fr/.

Bulf, C., Mathé, A.-C., Mithalal, J. (2014). Apprendre en géométrie, entre adaptation et acculturation. Langage et activité géométrique. *Spirale – Revue de recherches en éducation*, 54, 29–48.

Cousin-Fauconnet, A. (1995). *Enseigner la géométrie au collège. Un chemin pour la découverte progressive par l'élève*. Armand Colin, Paris.

Duval, R. (2005). Les conditions cognitives de l'apprentissage de la géométrie : développement de la visualisation, différenciation des raisonnements et coordination de leurs fonctionnements. *Annales de didactique et de sciences cognitives*, 10, 5–53.

Fregona, D. (1995). Les figures planes comme "milieu" dans l'enseignement de la géométrie ; interactions, contrats y transpositions didactiques. PhD Thesis, Université Bordeaux I.

Groupe Géométrie de l'IREM de Paris (2021). Enseigner la géométrie au Cycle 4. Comparer des triangles pour démontrer. Brochure IREM no. 100, 2nd edition. IREM de Paris.

Houdement, C. and Kuzniak, A. (2006). Paradigmes géométriques et enseignement de la géométrie. *Annales de didactique et de sciences cognitives*, 11, 175–193.

Keskessa, B., Perrin-Glorian, M.-J., Delplace, J.R. (2007). Géométrie plane et figures au Cycle 3. Une démarche pour élaborer des situations visant à favoriser une mobilité du regard sur les figures de géométrie. *Grand N*, 79, 33–60.

Laborde, C. (1990). L'enseignement de la géométrie en tant que terrain d'exploitation de phénomènes didactiques. *Recherches en didactique des mathématiques*, 9(3), 337–363.

Mathé, A.C., Barrier, T., Perrin-Glorian, M.J. (2020). *Enseigner la géométrie à l'école*. Collection Les Sciences de l'éducation aujourd'hui, Academia L'Harmattan, Paris.

Mathé, A.C., Maillot, V., Ribennes, J. (2021). Enjeux langagiers, situations de formulation et de validation en géométrie. Un exemple de travail autour du cercle en CE2. *Grand N*, 108, 27–57.

Mithalal, J. and Moulin, M. (2019). Teaching construction programs through a work on language phenomena. *Proceedings of the 11th Congress of the European Society for Research in Mathematics Education*, February 6–10, Utrecht.

Perrin-Glorian, M.-J. and Godin, M. (2014). De la restoration de figures géométriques avec des instruments vers leur caractérisation par des énoncés. *Math-école*, 222, 26–36.

Perrin-Glorian, M.-J. and Godin, M. (2018). Géométrie plane : pour une approche cohérente du début de l'école à la fin du collège [Online]. Available at: https://hal.archives-ouvertes.fr/hal-01660837v2/document.

Petitfour, E. (2015). Enseignement de la géométrie à des élèves en difficulté d'apprentissage : étude du processus d'accès à la géométrie d'élèves dyspraxiques visuo-spatiaux lors de la transition CM2-6ème. PhD Thesis, Université Paris Diderot.

Petitfour, E. (2017a). Outils théoriques d'analyse de l'action instrumentée, au service de l'étude de difficultés d'élèves dyspraxiques en géométrie. *Recherches en didactique des mathématiques*, 37(2–3), 247–288.

Petitfour, E. (2017b). Enseignement de la géométrie en fin de Cycle 3. Proposition d'un dispositif en dyade. *Petit x*, 103, 5–31.

Pluvinage, F. and Rauscher, J.C. (1986). La géométrie construite mise à l'essai. *Petit x*, 11, 5–36.

2

How to Operate the Didactic Variables of Figure Restoration Problems?

2.1. Introduction

This study focuses on the teaching and learning of plane geometry at the beginning of primary school. To this end, we propose to study a specific type of figure reproduction problem, developed by Perrin-Glorian and Godin (2014): Figure Restoration Problems.

This chapter synthesizes theoretical elements that have been engaged to address the question of choosing the values of the didactic variables for these problems. How to choose them to develop a learning sequence? How to promote geometric conceptualization in students at the beginning of primary school?

After having presented the theoretical elements underlying the definition of figure restoration problems, we develop which deal with the conceptualization and development of schemes associated with this type of problem. These elements thus allow us to present our approach and to justify the values of the didactic variables chosen to elaborate a family of problems that we will experiment with at the beginning of primary school.

2.2. Theoretical framework

2.2.1. *Studying a specific type of problem: figure restoration*

In situations involving the restoration of a figure, the student must understand the figure to be reproduced, that is, analyze the different geometric objects that

---

Chapter written by Karine VIÈQUE.

constitute it. This apprehension is not self-evident. Duval (2015) has shown that it requires the cognitive development of a certain geometric gaze to be brought to bear on the figures, which he calls "geometric visualization". Therefore, Perrin-Glorian and Godin (2014), in the continuity of Duval's work, propose to accompany this change of gaze towards figures through restoration problems, by which the mobility of the gaze towards figures is queried so as to successfully restore the figure.

The definition of a figure restoration problem is based on the framework of the theory of didactic situations outlined by Brousseau (1998): a model figure, as well as a part of the figure to be obtained – the starting point – being given (Figure 2.1), the task consists of completing the starting point in order to find the initial figure using the instruments at one's disposal. The didactic variables of a figure restoration problem concern the possible choices from these different elements. Among them, the choice of the instruments provided constitutes a key didactic variable for this type of problem. These can be geometric form templates, a non-graduated ruler, a set square or a compass. Also, a cost on the instruments can be a didactic variable used to allow the research and the identification of certain properties on the figures.

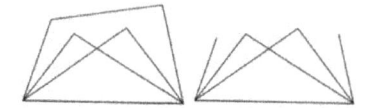

Figure 2.1. *Model figure and the beginning of the figure (Perrin-Glorian and Godin 2014, p. 30)*

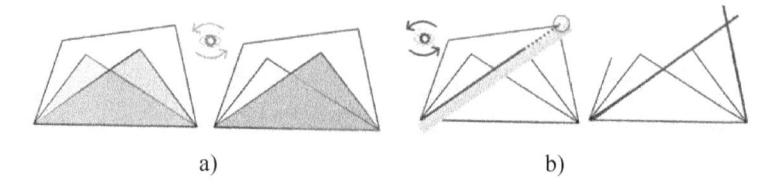

a) b)

Figure 2.2. *a) See the juxtapositions and superimpositions of figurative units. b) See the relationships between 1D and 0D figurative units of a figure (line and point visions). For a color version of this figure, see www.iste.co.uk/guille/tangible.zip*

We will illustrate how restoration problems are a means by which to act on the development of geometric visualization. In the example problem (Figure 2.1), identifying the relevant relationship to be taken in order to successfully restore the model figure requires changing our view of the figure several times. In particular, the student must mobilize both heuristic and mathematical ways of seeing a figure (Duval 2015). The heuristic way of seeing the figure (Figure 2.2(a)) consists of identifying juxtapositions and superimpositions of 2D figural units ("surfaces" view). The mathematical way of seeing the figure (Figure 2.2(b)) consists of identifying the relationships between 1D and 0D figural units of a figure (line and point visions).

Therefore, in our example that only uses a ruler, the success of the figure restoration relies on the identification of an alignment relation for a segment with a point, which requires the mobilization of a vision of the "lines" and "points". It is also a question of using the ruler, in its function of taking and carrying forward an alignment relation, for its function of tracing lines. The pupil must, indeed, know that in order to place the ruler, they need two points or a segment that has already been drawn: this is a "geometric use of instruments because it respects rules that implicitly refer to axioms, definitions or theorems of geometry" (Barrier et al. 2020, p. 55, *translated by author*). Therefore, the instrumental reproduction of figures can be a means of progressing towards conceptualization in geometry, "in a mutual development of construction techniques with instruments and of geometric concepts" (ibid. p. 56). At the beginning of primary school, however, drawing instruments (non-graduated ruler, set square, compass) are gradually introduced. Their geometric use is not self-evident; it has to be constructed. In this study, we seek to determine how to encourage the development of conceptualization in geometry by developing geometric visualization and by taking into consideration the geometric use of instruments.

2.2.2. *Studying the concepts involved in figure restoration problems*

We begin by considering the different semiotic representations of the concepts involved. According to Duval, "two representations are different when their contents are of a different nature, that is, they do not represent the same type of units (words, outlines)" (Duval 2005, p. 69, *translated by author*). Moreover, for a semiotic representation to give rise to a transformation, "it must be obtainable and modifiable by a production procedure, that is to say, a procedure that makes it possible to transform a representation that one gives oneself into another representation of the same kind" (ibid., p. 74, *translated by author*). Let us take as our example to support these remarks: Figure 2.2(a) illustrates looking at different vertices and sides of the subfigures composing the model figure, which can be obtained graphically by tracing the outline of geometric form templates. Figure 2.2(b) illustrates a look at segments, points and alignments between these segments and points, which can be obtained graphically by drawing straight lines and points of intersection with the ruler. Therefore, for the purposes of this study, we retain the importance of studying, for each new tracing instrument introduced in primary school, how the learning of new procedures can allow for the semiotic transformation of representations for the concepts of point, segment and line.

To study how to develop and operate the concepts involved, we use Vergnaud's definition of a concept. A concept "C" is a triplet of three sets "S, I, φ". The set "S" is the set of situations giving meaning to the concept. The set "φ" is "the set of linguistic and non-linguistic forms that make it possible to symbolically represent the concept, its properties, the situations and the processing procedures (the signifier)" (Vergnaud 1990, p. 145, *translated by author*). As mentioned above, we have just illustrated part of the content of this set by explaining certain registers of representation in play in Examples 2.2 and 2.3. Finally, set "I" is "the set of invariants upon which the operationality of the schemes (the signified) is based" (ibid., *translated by author*). A scheme is defined as "the invariant organization of conduct for a given class of situations" (ibid., p. 136, *translated by author*). Vergnaud also emphasizes that "there is no conceptualization without the construction of invariants, especially since it is a question of mastering a variety of situations. It is the invariants that will cut up the real, bring about conceptualization" (ibid., *translated by author*). The operative invariants are one of the components for the analytical definition of a scheme, given by Vergnaud. They are based on "categories and relations that make it possible to extract relevant information (concept-in-action), as well as propositions that are held to be true (theorems-in-action)" (Vergnaud 2002, p. 40, *translated by author*). This component of a scheme is closely linked to the rules of action, the information and control intakes. Indeed, the rules of action make it possible to relate actions to conditions (relevant information intake) and circumstances (goals to be achieved). They can be expressed in the form: "If ..., then". The study of rules of action thus makes it possible to take into consideration the cognitive functioning of the pupil, to grasp the importance of seeking how to act on the development of the "conscious decisions that make it possible to take into account the particular values of the didactic variables of the situation" (Vergnaud 1990, p. 137, *translated by author*).

For this study, we retain an interest in making explicit the rules of action and the theorems-in-action, upon which the operative invariants are based, in order to identify the values of relevant didactic variables in order to design different families of restoration problems.

Our intention is to act on the genesis of the schemes associated with figure restoration problems at the beginning of primary school. This condition seems to us necessary to favor the emergence of the geometrical reasoning upon which the adequacy of the actions performed with the instruments will be based, and thus to engage the conceptualization of geometry. We now present the values of the didactic variables of the first family of problems that we have designed.

2.3. Values of the didactic variables of the first problem family

2.3.1. *Values of the didactic variables for the "figure" and the "beginning of the figure"*

The work of Duval (2015) concerning the difficulties related to geometric visualization allows us to operate on the choice of the value to give to the didactic variable of the "figure". In order to promote the development of a heuristic way of seeing, we make the choice of elaborating the complex figures, that is, as being composed of simple figures (Figure 2.3). Students must learn to identify juxtapositions and superimpositions of subfigures to solve a problem. For example, they must learn to orient their gaze on the existing relationships between the sides of the subfigures that make up the complex figure.

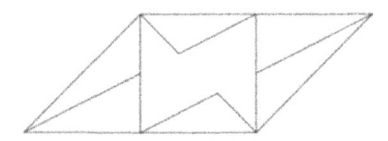

Figure 2.3. *A complex figure prepared for our experimentation*

To encourage the development of the mathematical way of seeing, the choice of the value of the didactic variable "the beginning of the figure" must act on the transition from an initial glance of the 2D units to a glance of the 1D and 0D units of the figures, that is, to lead the pupils to identify on figure segments, straight lines, points and existing relations between these objects. To do this, problem solving may require tracing straight lines not visible on the model figure, in order to obtain segments or points of intersection necessary to restore the figure. There are two levels of difficulty to consider here. From a "line" perspective, "the points are ends of lines or intersections of lines that we already have" (Barrier et al. 2020, p. 39, *translated by author*).

On the contrary, through a "vision point", a point can be the point of intersection for other lines than those present on the model figure. Taking into account these two levels of difficulty allow us to operate the values of the didactic variable "starting point": at the beginning of primary school, we choose to elaborate points for the beginning of the figure, for which the production of the missing points mobilizes the surface (Figure 2.4) and line (Figure 2.5) visions but not the "point" vision (Figure 2.6). We can now continue with the study of the values of the variable "instruments made available".

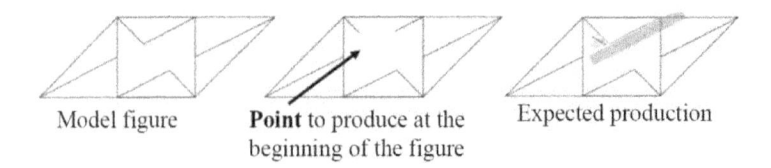

Model figure   Point to produce at the   Expected production
              beginning of the figure

Figure 2.4. *Example of mobilizing surface vision: production of a point as the common extremity of two segments present on the model figure*

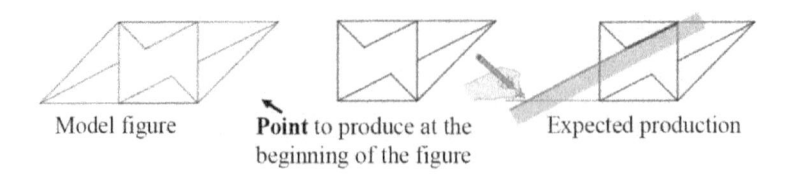

Model figure   Point to produce at the   Expected production
              beginning of the figure

Figure 2.5. *Example of mobilizing the line vision: production of a point as the intersection of the lines present on the model figure*

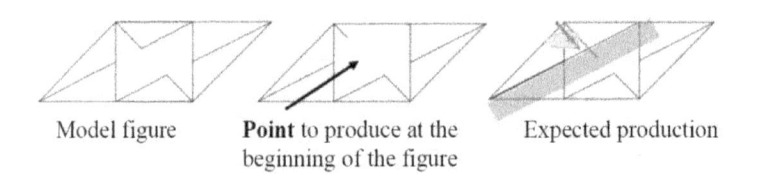

Model figure   Point to produce at the   Expected production
              beginning of the figure

Figure 2.6. *Example of mobilizing the point vision: production of a point as intersection of other lines than those present on the model figure*

2.3.2. *Value for the didactic variable "instruments made available"*

In order to act progressively on the conceptualization, we will analyze the existing relation between the evolution of the instrumented actions carried out by the pupils and that of the semiotic representations of the concepts towards point, segment and line. To do this, we begin by categorizing the different ways of graphically producing a point and a segment according to the instruments made available.

Therefore, with the use of a polygonal template or the ruler, a vertex point of a figure can be obtained as the common end of two segments. With the use of the ruler as the only instrument, a point can also be obtained as the intersection of two straight lines. With the use of the ruler and a length transfer line, a point can be obtained as the intersection of a straight line and a length measurement line. Using the ruler and the compass, the point can be obtained as the intersection of a straight line with a curved line. Finally, using just the compass, the point can be obtained as the intersection point of two curved lines.

This results in different ways of graphically producing a segment depending on the instruments available. With the use of a geometric form template, a side segment of a figure can be obtained by drawing a straight line along one of the sides of the geometric form template. This defines a side segment of a figure as well as a well-defined section of the figure outline. Using the ruler, we can obtain from a line supporting its direction, by identifying the two points located at the ends. This makes it possible to define a segment as a portion of a straight line delimited by two points, called the ends of the segment. With the use of the ruler and a length measurement, or with the use of the ruler and the compass, it can be obtained from a line supporting its direction, the data of one of its ends and the data of its length. With the use of the ruler and the compass, a segment can also be obtained from a line supporting its direction and its length. In these last two cases, a segment can be defined as a portion of a line of finite length.

These different ways of graphically producing a point and a segment allow us to categorize different families of problems. The resolution for each of our families of problems is based on the identification of a basic relevant relation to be taken and plotted with the help of an instrument. In this way, our approach aims at enriching the students' knowledge on the concepts of point, segment and line, by building upon the geometric property of each new instrument as it is introduced. For this study, we elaborate a first family of problems by fixing the value of the didactic variable "instruments made available" by making the ruler the only instrument available. Our objective is to lead the students to conceptualize the geometric notions associated with the learning of the geometric use of the ruler: to produce a segment as a portion of a straight line delimited by two points, to produce a point as the point of intersection of two straight lines, to identify and reproduce the alignment relations between segments, to extend segments and to draw straight lines using the ruler.

2.3.3. Rules of action and theorems-in-action associated with development on the geometrical usage of the ruler

We try to elaborate a first family of restoration problems by fixing the non-graduated ruler as the value for the didactic variable "instrument made available". The study of the linked schemes and their different components allows us to advance our approach. Indeed, in terms of signified, the "operative invariant" component of a scheme allows us to specify the rules of action and the theorems-in-action underlying the construction of signifiers.

A first rule of action linked to the geometric use of the ruler (Figure 2.7) can be expressed as follows: "If I have two points, then I can position the right edge of the ruler on these two points and draw a line connecting these two points, in order to represent the missing segment" (goal to be reached). It is based on the (implicit) use of two theorems-in-action: "A segment is carried by a straight line" and "A straight line passes through two points". This allows us to define a vertex of a figure as being a point (common extremity of two segments) and a side of a figure as being a segment (part of the outline of a figure bounded by two consecutive vertices of the figure).

Model figure   Beginning of the figure   Rule of action no. 1

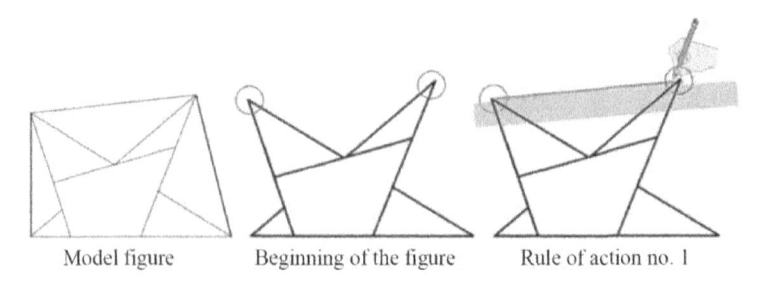

Figure 2.7. *Rule of action no. 1*

A second rule of action (Figure 2.8) can be expressed as follows: "If part of a segment is given, then I can position the right edge of the ruler on the segment's supporting line and extend it to represent the straight line on which the sought-after segment is then placed" (subgoal to be reached). It is based on the use of the theorem-in-action: "A segment is supported by a straight line". It allows us to define a segment as a portion of a straight line delimited by two points called the ends of the segment and a point at the intersection of two straight lines.

Model figure   Beginning of the figure   Rule of action no. 2

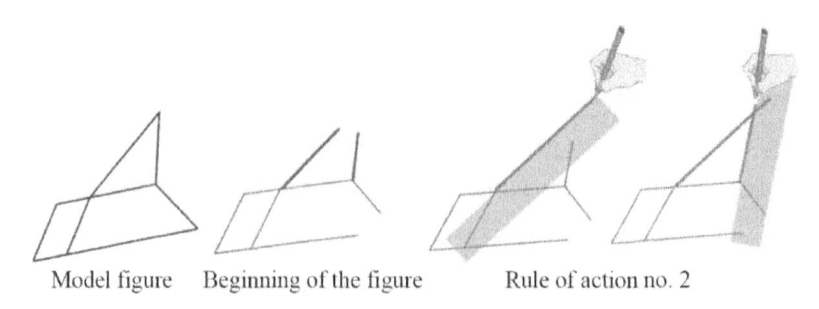

Figure 2.8. *Rule of action no. 2*

Finally, the third rule of action (Figure 2.9) can be expressed as follows: "If a segment is aligned with the one sought, then I can position the right edge of the ruler

How to Operate the Didactic Variables of Figure Restoration Problems? 43

on this segment and extend it to represent the straight line on which the new segment is placed" (subgoal to be achieved). This third rule of action is also based on the use of the theorem-in-action: "A segment is supported by a straight line". It allows for the construction of the alignment property between two segments, that is, between a segment and a point.

Model figure   Beginning of the figure   Rule of action no. 3

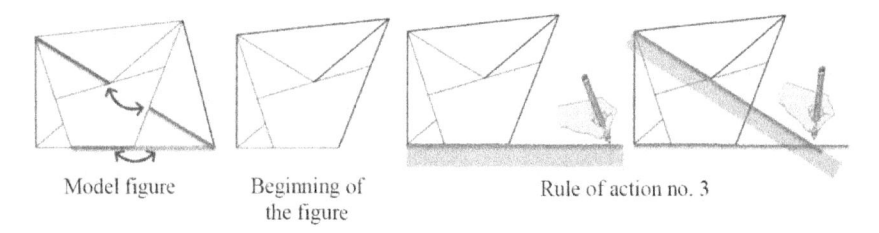

Figure 2.9. *Rule of action no. 3*

Therefore, in order to act on the conceptualization of the different objects at stake, our approach to finalize the elaboration of the beginning of figures is based on the relation between the expected instrumented actions and the conditions and circumstances, that is, the rules of action that need to be mobilized.

This allows us to elaborate a first family of problems that we will experiment with at the beginning of primary school. This family is composed of six figures (Figure 2.10). It will be integrated into a sequence aimed at learning the concepts of point, segment, straight line and the alignment of segments. This learning will be linked to the learning of the rules of action associated with the geometric use of the ruler.

Problem family	Problem 1	Problem 2	Problem 3	Problem 4	Problem 5	Problem 6
Model figure						
Beginning of the figure						
Rules of action mobilized	Rule of action no. 2	Rules of action no. 1 and no. 2	Rules of action no. 1, no. 2 and no. 3	Rules of action no. 1, no. 2 and no. 3	Rules of action no. 1, no. 2 and no. 3	Rules of action no. 1, no. 2 and no. 3

Figure 2.10. *Rules of action to be mobilized for the family of problems being developed*

2.4. Conclusion

At the beginning of this chapter, we raised the question as to the choice of values for didactic variables in order to develop restoration problems which are to be proposed at the beginning of primary school. This question seems to be an important consideration in order to determine how to jointly develop the geometric use of instruments and geometric concepts through the proposal of figure restoration problems.

Theoretical contributions have allowed us to adopt an approach that can identify the values to be given to didactic variables in order to propose a first family of problems for the purpose of figure restoration. The choices made aim at acting on the development of geometric conceptualization and visualization. The aim is for students to learn to see the vertex of a figure as a point (the point of intersection for two straight lines) and the side of a figure as a segment. The student must understand that a segment has a direction: it is supported by a straight line. They must also understand that they need to produce two points at the ends of a segment in order to trace it. The resolution of the family of problems is based on the identification of an alignment relationship between segments. Therefore, the choices made aim to encourage the construction of procedures related to the geometric use of the ruler, so as to initiate a certain transformation of representations for the concepts of point, segment and line.

The next part of our reflection concerns the ways in which we can make emerge theorems-in-action, which underlie the implementation of adequate actions. Indeed, we want to observe within students an effective adaptation of the schemes associated with the family of restoration problems that we have developed.

We also seek to determine the different types of didactic use situations that are necessarily articulated in our learning sequence, in order to lead students to construct the geometrical reasons by which the use of their instrumented actions is justified.

2.5. References

Barrier, T., Mathé, A.C., Perrin-Glorian, M.J. (2020). *Enseigner la géométrie élémentaire. Enjeux, ruptures et continuités*. Academia L'Harmattan, Paris.

Brousseau, G. (1998). *Théorie des situations didactiques.* La Pensée Sauvage, Grenoble.

Duval, R. (2005). Transformation de représentations sémiotiques et démarche de pensée en mathématiques. *XXXIIe Colloque de la COPIRELEM*, 67–89.

Duval, R. (2015). Figures et visualisation géométrique : "voir" en géométrie. *Du mot au concept*, 147–182.

Perrin-Glorian, M-J. and Godin, M. (2014). De la reproduction de figures géométriques avec des instruments à leur caractérisation par des énoncés. *Revue Math-école*, 222, 28–38.

Vergnaud, G. (1990). La théorie des champs conceptuels. *Recherche en didactique des mathématiques*, 10(2–3), 133–170.

Vergnaud, G. (2002). L'explication est-elle autre chose que la conceptualisation ? In *Expliquer et comprendre en sciences de l'éducation*, Saada-Robert, M. and Leutenegger, F. (eds). De Boeck Supérieur, Louvain-la-Neuve.

3

Early Geometric Learning in Kindergarten: Some Results from Collaborative Research

We present here some of the results of a collaborative research project (Morissette 2013) aimed at identifying the foundational elements of early geometric learning in kindergarten and, consequently, at providing teachers with the tools to help them better adopt this learning into their methods.

The various actors involved in this research – two teams of kindergarten teachers and their pupils, aged 3 to 5, a teacher trainer, a pedagogical advisor, a part-time teacher competent with digital technology and a teacher-researcher – shared the same problematic, namely: How to support kindergarten pupils to access their first geometric knowledge: Which kinds? Through which material? Through which problems? With which lexicon?

After recalling the questions that led us to this research topic and the theoretical clarifications we needed, we will use a few illustrative examples of the data collected so far, in order to bring out the first results and thus reveal some of the teachers' concerns in relation to their usual methods and practices, and their reactions when unusual situations and materials are proposed.

3.1. The emergence of the first questions

The materials on geometric shapes that circulate in kindergarten classes are often used for early-learning or leisure activities: the geometric knowledge that they could

---

Chapter written by Valentina CELI.

convey as such remains implicit. On the contrary, in other cases, only common shapes are used (square, triangle, rectangle, disk), and the focus is too quickly made their geometric properties. When the emphasis is placed on the importance of manipulation, a kind of "activism" prevails over the intellectual activity that could be aimed at (Caffieaux 2011): for example, in the case of an assembly of shapes, if the knowledge leading to success is not made conscious, the pupil will remain in the manipulation of shapes without benefiting from this activity. These observations led to the emergence of initial naive questions, namely: Is it possible to teach geometry in kindergarten? If it is possible, what geometric knowledge can be taught? What geometric knowledge can a pupil at this grade level learn? Assuming that the kindergarten is the place where the pupil can be guided towards their first geometrical learning, three aspects seem important to take into account: the materials one places into their hands, the problems that one proposes to them to treat with the help of these materials and the lexicon that they can/should learn during this geometrical activity. The concepts of artifact (Rabardel 1995) and semiotic potential (Bartolini-Bussi and Mariotti 2008) allow us to clarify our questions. Indeed, when in kindergarten class, the teacher and their pupils work with geometric shapes, they manipulate artifacts. The relationship that exists between the use of these artifacts and the underlying mathematical knowledge constitutes their semiotic potential. Our questions thus become: in the perspective of guiding pupils towards the first geometric learning, is the kindergarten teacher aware of the semiotic potential of the artifact they are placing in their hands? Are they aware that this work may correspond to a period of preparation for future geometric learning? At this stage, we felt the need to resort to other theoretical contributions that would allow us to shed light on these initial questions.

3.2. Theoretical insights

How does a child perceive geometric shapes? How can we help them to progress in the ways in which they perceive the shapes they manipulate and thereby guide them towards their first geometric learning? In this section, we propose some answers to these questions, which seem, to us, to be of primary importance when considering relevant access to this type of learning.

3.2.1. *Global understanding and visual perception of geometric shapes*

When speaking of *syncretic perception*, Claparède (1908) meant that the child perceives the world in a global way. Later, in relation to geometry, other authors confirmed this idea. For her *Casa dei Bambini*, at the beginning of the 20th century, Maria Montessori developed artifacts, geometric shapes made of wood, but she clearly discarded the introduction of properties, implying that it is first by its globality that a shape is perceived:

> One could have the idea of the square shape without knowing how to count to four and, consequently, without considering the number of sides and angles. The sides and angles are only abstractions, which do not exist by themselves; what exists is this piece of wood of a determined shape (Montessori 1970[1], p. 59, *translated by author*).

By comparing visual perception to language, Vygotskij (1980) also evokes the global character of this perception. Similarly, later on, and even more recently, other authors – Van Hiele (1959), Duval (1994), Rouche (1999) – confirm the idea that a child first recognizes shapes in a global manner. Notably, in defining five abstraction levels of geometric thinking, Van Hiele (ibid., p. 201) states that level 0 corresponds to the one wherein the learner recognizes shapes by their global aspect and classifies them exclusively in relation to each other: "A child recognizes a rectangle by its shape and a rectangle seems to be different from a square" (ibid., *translated by author*). Gentaz et al. (2009) show that, among the usual shapes, the triangle is the least well-known shape for children between the ages of 4 and 6 years, because non-similar copies of the same shape are not recognized in an equivalent way, and the triangle has an infinite number of such copies. This result then helps us to refute a belief shared (Vause 2011) by many teachers, namely, that "it is better to start the work of shape recognition with the triangle" (*translated by author*), because pupils in the early stages (3 years old) only know how to count to three. These teachers make this choice because their work on shapes focuses too quickly on the usual shapes and their geometric properties. They seem to ignore the fact that a child first grasps shapes in a global manner, which prevents them from guiding that child towards a progressive detachment from this spontaneous perception.

3.2.2. *Operative understanding and visual perception of geometric shapes*

When the pupil looks for a rectangle in an assortment of shapes, they must learn to recognize it independently of its size and position on the table: they understand it perceptually, and also, according to Duval (1994), *operatively*, according to its *mereological, optical and positional modifications*. Regarding the latter, Gentaz

---

[1] Montessori's *Pédagogie scientifique*, Volume 1, published in France in 1970, is a translation of *Il metodo della psicologia scientifica applicato all'educazione infantile nelle casa dei bambini*, of which there are several editions published between 1909 and 1950. Presumably, the work in French is translated from the English edition of 1912 or from the Italian edition of 1913 (see Celi et al. 2019, p. 37).

et al. (2009) show that the visual recognition of shapes is, among other things, determined by their number of axes of symmetry and by the orientation of the axes of symmetry with respect to the subject's body. This is therefore an important element to take into account in order to better identify pupils' difficulties when they approach this type of problem. This leads us to justify the difficulty that pupils have in recognizing the square, by confusing it with other shapes depending on its position. This is the case in the following exchange, where the teacher asks her pupils to name a square shape that she is holding in her hands by one end, in a non-prototypical position:

> Pupil 1: A diamond
> Teacher: A diamond?
> Pupil 2: A square
> Teacher: A square/how can you tell it is a square?
> Pupil 2: Because if I put my head like this [tilts his head so they can see the square shape in the prototypical position], it's a square
> Pupil 3: When I look like this it's a diamond and when I look like this [tilts head] it's a square.

The first pupil provides a term from non-geometric language. In response to the teacher's doubtful reaction, a second pupil provides the expected name, but to do so, they need to tilt their head in order to find the shape in a prototypical position (Figure 3.1). A third pupil considers that when their body changes position, it is not the same shape.

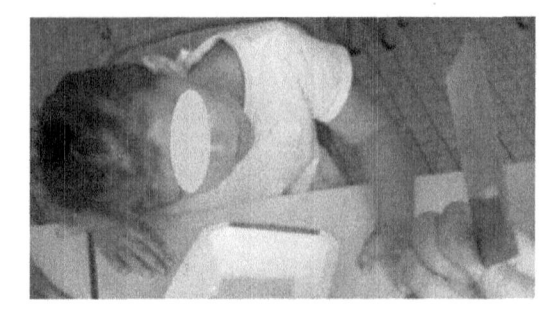

Figure 3.1. *A pupil tilts his head to recognize a square*

3.2.3. *Topological understanding and visual perception of geometric shapes*

According to the Piagetian approach (Piaget and Inhelder 1947), the representation of space is constructed by passing through three stages: during the first stage – known as topological space – the child does not easily distinguish what is curved from what is straight. This has led us to speak of topological understanding. When dealing with a shape classification problem, topological understanding often leads pupils to include in the category of triangles a portion of a disk and a convex quadrilateral with an angle close to 180° (the dark shapes in Figure 3.2).

Figure 3.2. *A shape classification problem*

3.2.4. *Haptic perception*

Montessori (1970) already emphasized the importance of associating the tactile-muscular sense in the *visual* sense, to promote the recognition of shapes:

> Undoubtedly, the association of the tactile-muscular sense with the visual sense helps a lot in the perception of the shapes and fixes the memory of it (*translated by author*).

Starting from the observation that a large part of fundamental school learning mobilizes only the *visual sensory modality* of young children, Gentaz et al. (2009) show today that, in an activity intended to prepare geometric learning, the *manual haptic modality*[2] allows children to better represent basic plane shapes:

> The introduction of haptic modality in figure recognition exercises and in the use of appropriate vocabulary would help children to better represent basic plane shapes thanks to its analytical processing and/or its double coding (visual and motor) (ibid., p. 29, *translated by author*).

---

[2] "*Haptic perception* results from stimulation of the skin through active exploratory movements of the hand coming into contact with objects. This is what happens when, for example, the hand and fingers follow the contour of an object to appreciate its shape" (Gentaz et al. 2009). This is what Montessori (1970) calls the *tactile-muscular sense*.

[...] manually perceiving a shape involves more analytical processing of information, unlike the visual modality involving more global processing (ibid., p. 31, *translated by author*).

In the search for the shape that fits exactly into the corresponding hollowed-out shape, a pupil may encounter failure. However, with the help of the teacher who invites them to "touch" its edge (Figure 3.3), they will be able to go beyond the global and topological understandings of the shape and grasp its properties. Haptic perception thus enables the child to better construct their first representations of basic plane shapes.

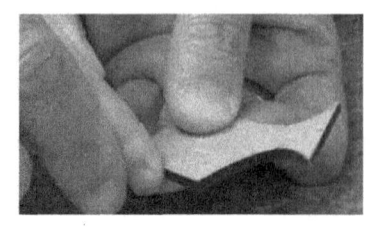

Figure 3.3. *With the help of the adult, the pupil's finger runs along the contour of the shape*[3]

3.2.5. Association of visual and haptic perceptions: towards a sequential understanding of geometric shapes

Through visual perception alone, all the properties of a shape are perceived simultaneously, whereas its association with haptic perception allows the pupil to go beyond global and topological understandings and to begin to understand shapes in a more analytical, *sequential* manner.

Duval (1994) speaks of sequential understanding in order to signify the order when constructing a shape, an order that depends on the geometric properties of the shape as well as on the instruments available for its construction. In the context of kindergarten, where visual and haptic perceptions play a fundamental role when a pupil manipulates shapes, we prefer to speak of *sequential understanding* so as to distinguish it from global understanding, one being more related to haptic perception and the other more attuned to visual perception:

---

3 The shape seen in this photograph (and later on in Figure 3.7) is part of an assortment of shapes made by Sylvia Coutat and Céline Vendeira-Maréchal (www.unige.ch/fapse/dimage/fr/recherche/reconnaissance-de-forme-geometrique/).

Tactile perception and visual perception are different when processing the different properties of an object. Indeed, in vision, all properties are perceived almost simultaneously, which is not the case in the haptic modality, due to the mode of exploration and motor incompatibilities rendering the perception very sequential (Fernandes and Vinter 2009, p. 410, *translated by author*).

This is how the child who looks at a shape and "touches" its outline combines visual and haptic perceptions: they are thus led to go beyond the global understanding of the shape, through a more analytical processing, towards a sequential understanding of it.

We recognize here that, when the pupil is encouraged to mobilize haptic perception, they understand the shapes they manipulate in a sequential manner: they are thus helped to modify the ways in which they look at them and to progressively move from a "surface" vision, where global understanding may suffice to a "surface contour" vision (Mathé et al. 2020, pp. 37–38). They thus undertake this process of dimensional deconstruction which will lead them to later consider the figures as being "made up of figurative units of smaller dimensions, put in relation by geometrical properties" (ibid., p. 36, *translated by author*).

3.3. The role of language in early geometric learning

As the haptic perception associated to visual perception leads to a more analytical treatment of the figure, the language by its nature does the same. This passage, in an excerpt from Vygotskij (1980, p. 54)[4], as such seems significant:

> Independent elements in a visual field are perceived simultaneously, in this sense visual perception is integral. Language, on the other hand, requires a series of consecutive processes. Each element is labeled separately [...] which makes language essentially analytic (*translated by author*).

If we consider haptic perception as one of the ways of acting on shapes, we find here the association of action, language and visual perception in the service of learning about shapes. In the French official texts, currently enforced in kindergarten, manipulation, action and language are also explicitly associated. We read in particular that:

---

4 This excerpt, translated by us, is taken from *Strumento e simbolo nelle sviluppo del bambino* (1930), one of the writings contained in a book published in Italy in 1980.

The approach to plane shapes […] is made through visual perception, manipulation and coordination of actions on objects. This approach is supported by language: it makes it possible to describe these objects and actions and promotes the identification of initial descriptive properties (MEN 2021, p. 16, *translated by author*).

We consider here the gestures associated with haptic perception, and also gestures such as *superimpose, juxtapose, flip, rotate*, etc. We will come back to these later.

3.3.1. *But which lexicon?*

The question thus arises as to the nature of the lexicon to be mobilized during this learning process, a concern shared by the teachers involved in the collaborative research. Their first concern being to characterize the usual polygonal shapes by the numbers of sides and vertices, the teachers wondered, in particular, about the pupils' use of the words peak or tip to designate a vertex: "Can we let them express themselves that way? Should we correct them right away? Which terms should be employed?"

We then took the opportunity to approach the question differently: is it relevant to start with the usual shapes and their properties? Taking into account the spontaneous perceptions of the pupils induced by the global and topological understandings, these exchanges led them to use, above all, a material that allowed the progressive introduction of a more or less adapted lexicon, while also accepting to share that which comes spontaneously from the pupils.

Figure 3.4. *Assortment of interlocking shapes: the lighter shapes fit into the black holder*

Early Geometric Learning in Kindergarten 55

After the first sessions and during a meeting, when the team of one school met with the other members of the project, one teacher talked about the vocabulary used by her pupils when dealing with a problem that exploited nestable shapes (Figure 3.4).

The activity of a pupil consisted of taking a full shape and, by visual perception alone, to find the corresponding hollow shape, the other pupils had to validate or invalidate their choice:

> Through validating, there is vocabulary which came out ... rounded, line, straight line, rounded, tip, peak. It's nice to do it this way, because then there is this vocabulary that comes out of them. For the moment we are there [...]. Curve was not said. They also didn't say corner. Square, round, triangle were said. One said rectangle. It's bigger, it's smaller. Afterwards I made them touch and they had to say: "Here, I can feel ..." a rounded, a straight line, a peak [...]. It's interesting because they can describe the sides [...], it's good because by touching they can feel the straight line, the peak.

Through this testimony, the group became aware of the vocabulary spontaneously used by pupils and the nuances that can exist between the terms used when they are looking at shapes or when they touch them, for example, in the case of *tip* and *peak* to designate the angular points of the shapes or *rounded* and *curved* to characterize their edges.

The same teacher then gave her pupils a new problem with the same materials. One pupil had to describe a full shape hidden in an opaque bag, relying only on haptic perception, and the other pupils had to identify the corresponding hollowed-out shape (Figure 3.5).

Figure 3.5. *Describe a shape by relying only on haptic perception*

Teacher: So what do you feel, Lilian?
Lilian: It spins/it spins a little bit more/it's sharp/there's a little bar

Teacher: So what shape do you think Lilian has? [A pupil points to a convex quadrilateral with a diagonal as the only axis of symmetry]. Do you think it's this one? [Pointing to this same quadrilateral] Lilian said it spins / is it spinning now?
Several pupils: No
Teacher: Can it be this one?
Several pupils: No [Lola points to a kind of sausage]
Teacher: It spins
Pupil: There's no peak
Teacher: You said it spins twice/does this one spin twice [pointing to the sausage]
Pupil: Yes, but there's no tip
Teacher: That's right, there's no tip [Several pupils now point to a circular half-crown] It turns [pointing to the half-crown] it turns twice [One pupil runs his finger along the rounded edges of this shape).
Lilian: It turns four times]
Teacher: Ah [he clarifies], Lilian says it turns four times/so, does this turn four times? [pointing to the half-crown] How many times does it turn? [A pupil, Cerise, points to the hollow square shape] Wait, Cerise/how many times does it turn on this shape, Jules? [pointing to the half-crown] One [while a pupil's finger touches one rounded edge] two [while the pupil's finger touches the other rounded edge] Two rounded edges, two curved
Multiple pupils: This one [pointing to the square shape]
Teacher: So Lilian said it turns four times/does this one turn four times? [pointing to the square shape] Lola?
Lola: Yes
Teacher: There, are there any lines that turn? Yes or no?
Several pupils including Lola: Yes
Teacher: Which ones? [Lola's finger runs along the edges of the square shape; other pupils do the same] Ah, because it feels like it's spinning like this [while her finger runs along the edges of the square shape] okay/but this line that I'm touching [repeatedly, their finger runs along a straight edge of the square shape] does it turn?
Several pupils: No
Teacher: How is it?
Several pupils: Straight
Teacher: It's a straight line [touching a second edge of the square shape] and there's another …
Several pupils: Straight
Teacher: [touching a third edge of the square shape] What is this one like?
Several pupils: Straight

Teacher: [touching a fourth edge of the square shape] What about this one?
Several pupils: Straight
Teacher: So, are there any lines that turn?
Several pupils: No
Teacher: Lilian told us that there were, that's right, there are lines that turn in your shape so it can't be this one [she points to the square shape].

Lola finally proposes a new shape (Figure 3.6) and, with the help of the teacher, checks that it has all the properties indicated by Lilian. The session then concludes with a validation: Lilian and Lola fit the full shape into the corresponding hollow shape.

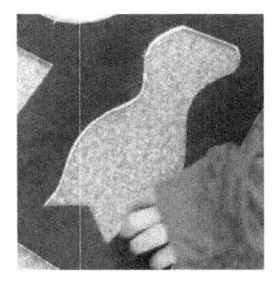

Figure 3.6. *The hollow shape corresponding to the full shape described by Lilian*

In these exchanges, the teacher lets the pupils' spontaneous vocabulary circulate. The latter seem to use the locution *it turns* in the direction which changes direction according to an angle (in the case of the square) or to designate a sinuous line (the shape described by Lilian) whereas the teacher seems to use it exclusively as a synonym for rounded. We note in passing that the locution *it turns* seems to refer to a dynamic image of movement, supported by haptic perception, whereas the term *rounded* seems to refer to a static image, supported by visual perception. Despite this misunderstanding, we note that an effort is made to consider more than one feature at a time based on those provided by Lilian – four rounded edges, two tips and one straight edge – which leads some pupils to refute proposals coming from their peers. It should also be noted that the material used and the problem posed make it possible to mobilize the various perceptions. The lexicon emerges from the sequential way in which the pupil understands the full shape that they are touching, which leads the other pupils to go beyond visual perception and global understanding towards a sequential understanding of the shape, by simultaneously mobilizing their haptic perception of the hollowed-out shapes. The inference of the material and the associated actions on the spontaneous vocabulary used by the pupils and accepted by the teacher is flagrant in the testimony of another teacher:

> We, it takes the edge off. You know, we talked about it in terms of the importance of vocabulary, and the fact that they're different shapes – at least I felt that way – we let them name things much more easily in their own words and through their own feelings. But here, I didn't find it important, if you will, the notion of precise geometric vocabulary. Of course, it was good that they could name, there you go, bumps and tips. A little girl in my class, when we used the octagon, as the tips were very small, said "No for me, they are not tips." We realized, however, that there was an awareness.
>
> There are many things that we left completely, without having the need to say to ourselves "We must give them the exact word" or modify theirs … It is true that it allows for setting a little distance on our part too.

When the teacher who is speaking here agreed to begin working with her pupils with non-usual geometric shapes, she realized that her requirement to introduce a precise geometric lexicon very early on became less important: by supporting them in this way, she enabled them to detach themselves from an excessive prevalence of global and topological understandings of the shapes being manipulated, placing more importance on the articulation between visual and haptic perceptions. Indeed, one can sometimes observe a learned lexicon that has been introduced too early, before the preliminary approach of the concept is relatively stabilized for the pupil, for example: Is it relevant to introduce the term *vertice* before having constructed the category of polygonal shapes?

3.3.2. *Verbal and gestural language*

Given the school level concerned, it is particularly interesting to analyze both verbal and gestural language. During a session, for example, in which a teacher proposed to her pupils in the first year (aged 3) a problem in which they had to learn to distinguish between a straight edge and a rounded edge, by means of a classification of shapes (Figure 3.7), we can clearly see that gesture becomes an indicator of knowledge during their interactions:

> Lena: It's the little balls [with a finger, she describes a circle in the air] it's balls
> Teacher: Ah because it's a ball [she takes a shape with a rounded edge in her hands]

Lena: Yes
Teacher: Does this look like a ball? [she points to polygonal shapes next to the round-edged shapes]
Several pupils: No
Teacher: Leo, is this straight [she takes a polygonal shape in her hand] with tips?
Several pupils: No
Teacher: What's this one like? [she touches the rounded edge of a shape]
Lena: This is round
Teacher: It is rounded, it is round; it is straight, with tips like here. [she asks, picking up a polygonal shape] What's this like? [she touches the rounded edge of the other shape] It's?
Lena: It's round [with one finger, she describes a circle in the air]
Teacher: It's rounded/so here [she goes back to the polygonal shapes] is it rounded?
Several pupils: No, no, no!
Teacher: So here [she points to two shapes with mixed edges] what's going on here?
Lena: This is the hat like the hat it's rounded
Teacher: Matteo [showing him a disc], what is this shape like? Does it have tips or is it rounded?
Matteo: Yes [with one finger he makes circles in the air]
Teacher: What does that mean? It means that it's [Matteo continues to draw circles in the air] rounded right / do you put it with the rounded shapes, Matteo? [Matteo puts the shape correctly in the pile of shapes with rounded edges].

Figure 3.7. *Polygonal shapes, rounded edge shapes and mixed edge shapes*

According to the typology of gestures established by Roth and Lawness (2002), we recognize here an *iconic gesture*, describing characteristic features of a concrete object, evoked in the conversation. Matteo is able to express his understanding of the concept of rounded through a gesture before being able to use the term. Lena, on the contrary, is able to articulate gestures and terms: she spontaneously names the *ball* first, making circles in the air, and then quite quickly incorporates the term *rounded* introduced by the teacher, while later returning to the spontaneous term *hat*, accompanied by a gesture on her head evoking a rounded shape. These spontaneous gestures and terms will be used by the teacher as personal signs to lead the pupils to a *mathematical* sign, in this case, the rounded edge (Bartolini-Bussi and Mariotti 2008). It should be noted, however, that she does not yet seem to have realized that a shape with rounded edges could also have *tips*: this is why she associates the latter exclusively with the vertices of polygonal shapes. Therefore, she passes very quickly over the shapes with mixed edges, which are, however, part of the shapes to be studied with the pupils.

This is undoubtedly due to the fact that, before agreeing to use non-usual shapes with her pupils, she worked mainly with triangles, squares, rectangles and disks, the question of *tips* associated only with polygonal shapes.

3.4. Assembling shapes

The various problems thought up in this research – and partially tested to date – involve four types of materials (Celi and Semmarty 2018; Celi 2019, Celi 2021): nestable shapes; shape sets (to make categories, for example); shape assemblies; and templates and stencils (to move from manipulation to graphing). In the following sections, we focus on problems encountered when using shape assemblies. We first study the possible deviations when proposing free assemblies with shapes from albums and then focus on the problem of assembling two isometric triangles, highlighting the difficulties encountered by a teacher when identifying the semiotic potential of this material, as well as the difficulties of some pupils grasping it; we end this section by setting out the questions we are currently asking ourselves about this type of problem and about the most appropriate way of bringing out the knowledge needed to handle it successfully.

3.4.1. *Free assembly of shapes*

At the beginning of our collaborative research, one of the teams of teachers involved envisaged a progression ending with the making of a scrapbook (inventing a story): these teachers built their sequence on geometric shapes around this objective and, for this, they also decided to start reading from the scrapbook. The shapes from this book were used in various sessions, one of which specifically focused on making free assemblies (Figure 3.8).

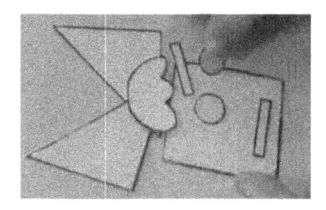

Figure 3.8. *An example of a free assembly made by a pupil*

The stated objective of the session was geometric shapes. However, the following excerpt shows that when the teacher asks a pupil to describe her assembly, their language interactions are exclusively about body parts, without any reference to geometric shapes or their properties:

> Teacher: What did you make for us?
> Pupil: A fox
> Teacher: A fox, okay, what did you put here?
> Pupil: The ears [she points to the two triangular shapes]
> Teacher: And why did you choose these ears? What's wrong with these ears?
> Pupil: They're big
> Teacher: They're big and then they're on top. They're ...
> Pupil: Pointy
> Teacher: Pointy. And they look like pointy fox ears, so why did you pick that for us? What did you put there? [points to a small rectangular shape]
> Pupil: A mouth
> Teacher: Why did you choose this shape for the mouth?
> Pupil: Because it scares people
> Teacher: Because it's / how is that mouth?
> Pupil: Lying down
> Teacher: Lying down? Right.

The word *pointy*, pronounced by the pupil, could mean the vertices of a geometric shape, here the triangle. But the child undoubtedly uses it in its common meaning; the teacher's remark: "Pointy. And they look like pointy fox ears" is also described through this common meaning. The term *lying down*, still used by the pupil, is borrowed from the vocabulary specific to the exercises of graphism, where the *vertical* and the *horizontal* are often expressed by the terms *standing* and *lying down*; the child then indicates the position of the *mouth*, not its nature.

Finally, when interacting with this pupil, the teacher fails to properly bring about geometric learning, such as specific vocabulary and the properties of shapes. In addition, the material in this pupil's hands – plastic-coated cardboard – is not suitable for haptic perception to encourage a sequential understanding of shapes. The teacher and the pupil remain focused on a visual perception.

These observations and the analyses which followed them, directed us towards the proposal of another type of assembly problem: to assemble typical geometrical shapes to reproduce typical geometrical shapes, in order to elicit an *iconic visualization* of shapes (Duval 2005), as well as mental images and lexicon which do not belong to the geometrical domain (houses, animals, parts of the body, etc.).

3.4.2. Assembling triangles

When the pupil engages in a problem of reproduction by assembling shapes[5], it is not enough for them to recognize the pieces globally, and they must be interested by comparison in the properties of the manipulated shapes: in the possible alignment of edges, in the *juxtaposition* of two pieces according to their edges, in the way that *corners* are juxtaposed.

This type of problem helps the pupil to go beyond his spontaneous perceptive understandings – global and topological – to move towards a sequential understanding of shapes through the analysis of the edges that constitute them.

Here, it is also the operative understanding of shapes, in its *mereological* modifications, that is involved (Duval 1994).

Several gestures can help the pupil to operate these modifications:

– *superimpose* the starting shapes in order to discover or verify their isometry, and in particular, the equality of the lengths of the homologous edges;

– *juxtapose* the starting shapes according to the various edges, to discover or verify the equality of the lengths of the homologous edges so as to arrange them appropriately;

---

5 In the programs currently in operation in France, the task of reproducing assemblies of shapes from a model is one of the expectations at the end of Cycle 1 (3–5 years). In the 2019 national assessments, for pupils in the Cours préparatoire (CP, 6 years old) and the Cours élémentaire 1 (CE1, 7 years old), it is proposed to deal with assembly problems using visual perception only, with the stated aim of assessing the ease with which the pupil mobilizes their spatial knowledge in order to perceptually identify certain geometric properties. The proposed problems seem to be psycho-technical tests: two isometric shapes are given, and the pupil has to circle the shape obtained by assembling them, from a choice of four options.

– *rotate* (rotation) or *flip* (indirect isometry) a shape in order to match it properly with another one.

Verbal exchanges around these gestures remain important here, as they can help the pupil associate mental images with the geometric knowledge that will be addressed later in school. Let us look now at the assembly of two isometric triangles:

– if the triangles are scalene, we get three parallelograms, two concave quadrilaterals and one convex quadrilateral, each with a diagonal as the only axis of symmetry;

– if the triangles are isosceles, we get a rhombus and a convex quadrilateral with a diagonal as the only axis of symmetry and a parallelogram;

– if the triangles are equilateral, we get a rhombus;

– if the triangles are right-angled, we get two isosceles triangles, a convex quadrilateral with a diagonal as the only axis of symmetry, a rectangle and two parallelograms;

– if the triangles are right-angled isosceles, we get a square and an isosceles triangle.

The way in which the triangles are placed on the table will considerably influence the assemblies that can be made. Let us consider the case where we have two right-angled triangles and we do not allow ourselves to use indirect isometry (flipping over a shape). If the two shapes are placed as in Figure 3.9, on the left, it is possible to obtain only two triangles and a convex quadrilateral with a diagonal as the only axis of symmetry; if they are placed as per Figure 3.9, on the right, it is possible to obtain only two parallelograms and a rectangle.

Figure 3.9. *Two ways of placing two triangular shapes on a table*

In addition, the rectangle and triangles could be identified more easily than parallelograms and the convex quadrilateral with a diagonal as the only axis of symmetry. The latter are less familiar to younger pupils.

The shapes to be made could be provided and used for different purposes. We could allow the assembly to be made on the given shape, which would require the pupil to properly *juxtapose* the two triangles and also to properly *superimpose* them onto the model provided. Alternatively, the pupil could be allowed to make the assembly next to the given shape, and the final product could be validated by visual recognition or by *superimposition* – if the shapes are at the same scale – but it is then important that the pupil has learned to have control over the actions performed in order to succeed.

It would also be possible to provide an outline of the shapes to be made, which could be used as a means of realization or validation; the superposition would be done by making the edges of the shape coincide with the graphic traces. In any case, it would be necessary to provide sufficiently rigid shapes so that the *superpimposition* is done easily. The various gestures important in the treatment of this problem: the teacher could check with their pupils that the triangles provided are isometric, that they are "the same", which would lead to the use of *superimposition* and, if necessary, of *flipping over*.

In late 2019, teachers involved in the research agreed to offer their pupils a shape reproduction problem using two isometric right-angled triangles. We focus here on one of these teachers, taking into account the data collected during the observed session and during the interview that took place later. The session observed consists of three phases: presentation of the material, the pupils' activity and then a pooling of the results. Five, varied and complex, educational materials are presented (Figure 3.10):

– the models of the six shapes to be reproduced, in colored and laminated cardboard, are posted on the board and available to all the pupils;

– the six shapes on transparent paper are also available to the pupils (these are the outlines of the six shapes);

– a *self-evaluation grid* and star-shaped stickers, two triangular shapes (these are isometric right-angled triangles) made of white cardboard and laminated are distributed to each pupil and will be used to make the assemblies.

During the interview, the teacher justifies her choices as follows:

> I had planned to use colored shapes because I told myself during the manipulation time, it would be easier to designate them, as much for them as for me [...] I told myself that if they were embarrassed or handicapped by vocabulary at least they could show these to us easily, designate them easily [...] the same shapes on a transparent support with the border, so that, it was easy to check, and which allowed me to

see whether they could manage to handle the transparency, that gave the possibility of positioning above or below and using them to differentiate a little [...] each one had a logbook (self-evaluation grid) to note their successes. With the small stars, we validated their success and that allowed, at the same time, for the teacher to not lose themself and to know what each one was trying, what remained to be figured out, or not, and which also allows the pupil to know what they made, what remains for them to figure out, and which is possibly playful, they like things that makes them the sheriff [...] and each child had two superimposable white triangles.

Figure 3.10. *Materials made available to pupils. For a color version of this figure, see www.iste.co.uk/guille/tangible.zip*

During the presentation to the pupils, the teacher very quickly mentions the way in which the pupils must or can articulate these various materials and does not realize that this corresponds to a large number of subtasks to be accomplished, not all of which are necessarily linked to the learning objective. The two triangles, which make it possible to carry out the proposed assemblies, are not named, the teacher presents them very quickly, she does not dwell on how to use them or on their common properties – for example, by superimposing them – although highlighting the fact that the two triangles are isometric is important, she does not:

> I am going to give everyone these two shapes, they are both the same. You can move these shapes around, change their position, turn them around, however you want to; you can do whatever you want with them.

Invited to express themselves about the six shapes to be reproduced, the pupils focus cognitively more on their colors, on their material of manufacture, than on their name and properties. The teacher engages with the pupils about two shapes, for which the recognition is quickly done and only by visual perception:

> Teacher: What do you think of these shapes?
> Pupil 1: They have colors
> Pupil 2: You can see plastic around them
> Pupil 3: You see blue, red, black
> Teacher: But if we talk about the shape, like when we go to touch them, what would we say? [she points to a yellow rectangular shape]
> Pupil 3: That's a square
> Teacher: This is a square?
> Pupil 4: A rectangle
> Teacher: Why?
> Pupil 5: It's a little bit longer
> Teacher: It's a little bit long, it's looonger [she next shows a red triangular shape]
> Pupil 2: It's a triangle.
> Teacher: A triangle, but how do we know that?
> Pupil 3: Because it has tips
> Teacher: It has tips [by touching it]

The excessive packaging of the situation, chosen perhaps to make it more attractive and motivating for the pupils, requires a long presentation and thus does not seem to favor highlighting essential information for its implementation. In fact, during the interview, the teacher admitted that she had not been able to anticipate difficulties – due to the situation itself or to the choice of presentation – that emerged during the session.

The pupils finally deal with the problem by encountering several pitfalls (Figure 3.11), and it is while they are doing the activity that the teacher becomes aware of these: *juxtaposing* sides of the same length and *flipping over* the shapes are two gestures that the pupils did not know how to produce spontaneously; *validating* the assembly with the help of the shapes on a transparent support is not as simple as she initially thought it would be.

The pupils do not seem to be aware that the two shapes are isometric, that their edges are "the same", and that in fact it is this property that allows them to successfully achieve the task at hand.

When they succeed and try to use the shapes on the transparent support as a means of validation, they are confronted with two types of difficulties: first, because of the material used for the pieces they are manipulating, and second, due to a lack of awareness of the strategies needed to succeed. If they try to place the support on the assembly they have made, they cannot make them coincide correctly because "everything moves". Conversely, when they try to place the completed assembly on the transparent support, they are faced with a new problem because they have to reconstitute the assembly on the support; however, they have forgotten how they made it, since the *juxtaposition* of the "same" edges is a gesture that they have not yet become aware of.

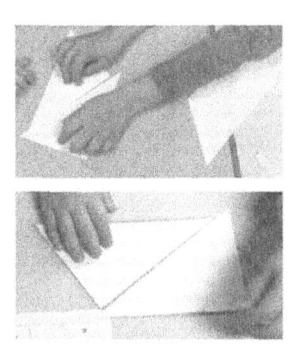

Figure 3.11. *Difficulties encountered by pupils to obtain a given shape and to use a transparent support*

During the collective review session, the first problem is dealt with relatively quickly:

> With these two shapes that you could turn and flip, you made a lot of other shapes. What shapes did you make?

The interactions between the teacher and her pupils do not concern the strategies adopted, the gestures for success, and the difficulties encountered, only the final product: the six shapes that can be made with the two triangles, their names and their properties. Far from considering it uninteresting, we nevertheless want to point out here that the initial problem and the geometrical learning it entails have been discarded; the realization of the product dominates at the expense of the intellectual activity it aims for. The proposed assembly problem favors the dissociation of the edge and the shape; the teacher seems to have difficulty identifying the mathematical knowledge at stake and the cognitive advances necessary to acquire them.

When the pupils succeeded, they did so most often "by chance". During the collective phase, the teacher does not shift the focus from the result of the pupils' activities to the way in which they achieved this result. She fails to examine with them how they might have arrived at the right solution (Cèbe et al. 2004). Language seems to be a communication tool here, not a tool for reflection and learning: the absence of language interactions about the problem posed does not allow each pupil to take control of their "action" either.

3.5. Gestures to learn

The analysis of the session on the assembly of two triangles imparted a lot to the various actors involved in this research. In terms of implementation, the teachers became aware of the importance of gradually bringing out various gestures through various problems: *superimposing* shapes to discover or verify their isometry, careful to note that the homologous sides coincide; *flipping over* a shape in order to arrange it correctly with another; *juxtaposing* shapes to discover or verify that the sides are, or are not, the same length. The gesture of *juxtaposition* is at the heart of the problems of assembling shapes and favors the distinction between the edge and the surface, by decentering the observation of the surface towards the edges of this surface, and thus initiating the first steps of a dimensional deconstruction. The gesture of *flipping over* a shape can emerge when working with problems involving interlocking shapes, which the teachers had not yet introduced in the previous year when working with this type of problem. For the *superimposition* of shapes, in order to check their isometry, we proposed some problems. For example, the one where, in a given set of shapes (Figure 3.12), the pupil has to find the same shape as the one in their hand: *superimposition* emerges as a means of validation.

Figure 3.12. *Find in an assortment of shapes the one that is isometric to the black triangular shape*

One of the teachers, realizing the importance of having her pupils learn these gestures, nevertheless encountered a completely different problem when the question asked required this gesture as an answer. She handed out two isometric triangles to each of her pupils, had them name them, and then asked, "What are these triangles like?" In response to the silence that arose, she then asked increasingly explicit questions:

Teacher: If we take your two triangles, do they look the same?

Pupil: Yes
Teacher: How much? What can you tell me/are they different or are they?
Pupil: They are the same
Teacher: How can you tell they are the same?

It is clear that the gesture in question – *superimposing* the two triangles – is not spontaneous. The pupils seem to be satisfied with a global understanding, possibly by placing them properly next to each other with the same orientation. However, they do not spontaneously think about *superimposition*. When the teacher encourages them to *move* the pieces, the pupils only allow themselves to move them within the plane of the table, without lifting them up off of it; she will eventually suggest that they put them "one on top of the other" and draw their attention to the fact that the homologous edges coincide. The treatment of gestures such as *superimposition, juxtaposition* and *flipping over* of shapes according to appropriately chosen problems is still the subject of ongoing discussion within the group.

Necessary for early geometric learning, we are thinking more about the mathematical knowledge evoked by these gestures and how to bring them out explicitly in a classroom setting. We have also realized that the pupils benefit from these learning objects themselves. Other problems will be experimented with, and the consideration of these gestures will probably be envisaged in the sequence of problems that we are currently developing.

3.6. Conclusion

Visually recognizing shapes and naming them is the usual culture of geometry teaching from the beginning of kindergarten. According to our observations, it is the main objective of many teachers. According to the elements that we have just exposed in this chapter, the taking into account of haptic perception is necessary to help the pupil to go beyond their spontaneous understandings of shapes, and thus understand them in a sequential way, by characterizing them by the nature of their edges.

Beyond the global recognition of shapes and their designations, this work constitutes one of the foundations of early geometric learning and prepares the ground for successive learning where, for example, the straight edges of a polygon will be identified as *sides*, the tips as *vertices*.

We have learned a lot from observing pupils doing activities: the ability to share a spontaneous vocabulary; the confidence they have in debating the nature of shapes; the pitfalls they encounter; and the ways to help them overcome these. Nevertheless, we would like to conclude here by focusing on a few orientations concerning teaching practices.

What about the educational practices of the teachers involved in collaborative research? While we have observed a tangible evolution of some professional gestures, certain resistances are still observable.

As multi-skilled teachers, it is quite legitimate for them to want to weave links between the disciplines and to tackle a more general theme that resonates with some of them: this was notably the case for teachers who began their sequence by exploiting a scrapbook, aiming to create an original scrapbook using various geometric shapes. When helped to realize that this choice was counterproductive in relation to the mathematical learning objectives being sought, these teachers agreed to work on shapes for their own sake, and no longer to include them in a project where geometric learning, although not absent, risked being marginalized and instrumentalized in relation to the final challenge. The problems proposed to the pupils increasingly take into account the articulation and alternation between visual and haptic perceptions and certain gestures – including *superimposing*, *juxtaposing* and *flipping over* shapes – which are now recognized as intrinsic to preliminary geometric learning.

The teachers also became aware that the nature of the artifact induces the nature of the language interactions by temporarily accepting the pupils' spontaneous vocabulary. However, the difficulty in independently questioning the semiotic potential of the artifacts at play and the problems posed is still omnipresent.

What geometric knowledge underlies the use of these artifacts? What is at stake in the problems posed?

The teachers still seem to have difficulty identifying the initial cognitive learning objectives and, consequently, to envisage the progression for *relevant* problems.

More generally, with regard to the theme of geometric shapes in the kindergarten classroom, what teaching practices would we like to see? Wouldn't we want an enlightened teacher to rely on a relevant analysis of the semiotic potential of the artifacts at play? That these pose real problems to the pupils? That these help pupils to become aware of their learning through language interactions? That these give priority to pupil learning over the communication of their successes? On this last point, the comment of a teacher who, in one of the progress reports of the research period, wrote:

Last year, we took pictures of the pupils' activities, and we shared them with families through the notebook. Pupils were able to tell their parents about what they had done in class as part of the project. This year we did not communicate with families about geometric activities as systematically, as our work focused on manipulatives and language activities in the classroom.

More broadly speaking, if teaching practices regarding geometric learning are limited today, might this not be better approached by querying the practice of evaluating learning? Do we only teach that which is easy to evaluate, that which we think we know how to evaluate? This brings us back to the question of what is at stake by this type of learning, in kindergarten and indeed beyond.

3.7. References

Bartolini-Bussi, M. and Mariotti, M.A. (2008). Semiotic mediation in the mathematics classroom: Artifacts and signs after a Vygotskian perspective. In *Handbook of International Research in Mathematics Education*, 2nd edition, English, L.D. and Kirshner, D. (eds). Routledge, New York.

Caffieaux, C. (2011). *Faire la classe à l'école maternelle*. De Boeck, Louvain-la-Neuve.

Cèbe, S., Paour, J.-L., Goigoux, R. (2004). *Catégo*. Hatier, Paris.

Celi, V. (2021). N'oublions pas la géométrie. *Cahiers Pédagogiques*, 573, 28–29.

Celi, V. and Semmarty, M. (2018). Pré-apprentissages géométriques à l'école maternelle [Online]. Available at: https://public.weconext.eu/academie-sciences/2018-12-12/video_id_005/index.html [Accessed 12 December 2021].

Celi, V., Coutat, S., Vendeira-Maréchal, C. (2019). Travailler avec les formes en maternelle : premiers pas vers des connaissances géométriques ? *Actes du 45e colloque de la Copirelem*. ARPEME, Paris, 35–55.

Claparède, É. (1908). La perception visuelle et la fonction syncrétique chez l'enfant. *Archives de Psychologie*, 7, 195.

Duval, R. (1994). Les différents fonctionnements d'une figure dans une démarche géométrique. *Repères-IREM*, 17, 121–138.

Duval, R. (2005). Les conditions cognitives de l'apprentissage de la géométrie : développement de la visualisation, différenciation des raisonnements et coordination de leurs fonctionnements. *Annales de didactique et de sciences cognitives*, 10, 5–53.

Fernandes, M. and Vinter, A. (2009). Développement des représentations graphiques réalisées par des enfants à partir d'une exploration tactile ou visuelle de formes bidimensionnelles. *L'année psychologique*, 109, 407–429.

Gentaz, É., Bara, F., Palluel-Germain, R., Pinet, L., Hillairet De Boisferon, A. (2009). Apports de la modalité haptique manuelle dans les apprentissages scolaires. *Cognito*, 3(3), 1–38.

Mathé, A.-C., Barrier, T., Perrin-Glorian, M.-J. (2020). *Enseigner la géométrie élémentaire. Enjeux, ruptures et continuités*. Academia L'Harmattan, Paris.

MEN (2021). Programmes de l'école maternelle. *Bulletin officiel spécial*, 2, 24 June 2021.

Montessori, M. (1970). *Pédagogie scientifique*, Volume 1. ESF, Paris.

Morissette, J. (2013). Recherche-action et recherche collaborative : quel rapport aux savoirs et à la production de savoirs ? *Nouvelles pratiques sociales*, 25(2), 35–49.

Piaget, J. and Inhelder, B. (1947). *La représentation de l'espace chez l'enfant*. Presses Universitaires de France, Paris.

Rabardel, P. (1995). *Les hommes et les technologies, une approche cognitive des instruments contemporains*. Armand Colin, Paris.

Roth, W.-M. and Lawness, D. (2002). Scientific investigation, metaphorical gestures and the emergence of abstract scientific concepts. *Learning and Instruction*, 12, 285–304.

Rouche, N. (1999). *Formes et mouvements. Perspectives pour l'enseignement de la géométrie*. CREM, Strasbourg.

Van Hiele, P.-M. (1959). La pensée de l'enfant et la géométrie. *Bulletin de l'APMEP*, 198, 199–205.

Vause, I. (2011). Des pratiques aux connaissances pédagogiques des enseignants : les sources et les modes de construction de la connaissance ouvragée. PhD Thesis, Université catholique de Louvain, Louvain-la-Neuve.

Vygotskij, L.S. (1980). *Il processo cognitivo*. Bollati Boringhieri, Turin.

4

Using Coding to Introduce Geometric Properties in Primary School

The production of writing in a classroom context participates in the conceptualization and structuring of thought (Verin 1995). Geometry is introduced in primary school through manipulation, following which the introduction of geometric tools participates in the progressive theorization of geometry. The geometric properties of figures appear with the premises for this theoretical geometry. This chapter proposes a reflection on the use of coding in primary school to accompany the introduction of certain geometric properties. After a brief clarification of what we call coding, two examples illustrate possible applications of coding in geometry with 8- to 12-year-old pupils. A discussion of the contributions and limitations of the two examples concludes this chapter.

4.1. Coding in geometry

The power of mathematics lies in its capacity for abstraction and the production of a specific language in conjunction with the development of mathematical theories. One of the specificities of this language rests on the coding of certain concepts. For example, signs are used to designate the geometric properties of a figure on a drawing, such as "⌐" to indicate two perpendicular lines.

According to Houdement and Petitfour (2018), the sign exists according to an interpretation that we can make of it and a referent to which it refers. If we define the geometric figure as a referent (in the sense of Vergnaud's triplet [1990]), a mathematical, theoretical object, which is defined by its properties, a representation of the figure in the graphical space can be a drawing. According to Laborde (2004),

---

Chapter written by Sylvia COUTAT.

this last one has two functions, namely that of the representation of an object within the tangible space, as well as that of the representation of a geometric figure. The drawing is not necessarily faithful to the figure that it represents. For example, it can show more properties than the figure contains (we are referring here to the triangle, which is often represented as an isosceles or even equilateral triangle). According to Salin (2003), the use of coding in geometry makes it possible to show that which is true in the representation (the drawing), and to indicate the properties of the figure. The coded drawing encompasses all the drawings that could illustrate the figure. As such, the strength of the coding lies in the encompassing of all representations of the target concept. Let us once again take the example of perpendicularity. At school, pupils are confronted with this concept very early on, for example, the recognition of the square without the property of perpendicularity being made explicit. It is only in the second half of primary school that this property becomes explicit, with the introduction of instrumented geometry (Houdement and Kuzniak 1998), and that coding can eventually be introduced. We will use the term "coding" to designate the set of graphical signs used to make explicit one or more geometric properties.

Different situations justify the use of coding. Coding translates a property of a drawing given in a discursive register or observed in a manipulable form. The use of coding allows us to attest the use of an instrument in a construction and thus state a specific property. It can make visible an invariance during a deformation of the Cabri-drawing (Laborde and Capponi 1994). Coding can be used in communication tasks, or in descriptions, to make specific properties explicit in an economical way. Finally, when it is associated with a drawing, it refers to a geometrical figure defined by its properties. There is a different coding that depends on the register or property to be encoded. In the case of conventional coding in the graphical register, it is possible to code the equality of measure (of segment length or angle), perpendicularity and parallelism of two lines. The use of physical objects is common among pupils when introducing the practice of geometry through manipulable shapes (Vendeira-Maréchal and Coutat 2017). These manipulable shapes give way to drawings and an engagement in graphical space.

However, as we have seen, these drawings have a double function. They are considered as "models" for the manipulable shapes used by pupils and the representation of geometric objects used by experts. How can we make pupils aware of the fact that the drawing is not the actual figure? We propose two situations that illustrate the use of coding to show that the drawing goes beyond its function as a model for representing figures in tangible space, and instead becomes a representation of geometric objects by representing their properties.

4.2. Two examples of communication activities requiring the use of coding

The first example is an extension of Vendeira-Maréchal and Coutat's (2017) research, and proposes the use of coding for a first abstraction of characteristics to introduce geometric properties. The second example takes up the work of Coutat (2018) around the use of dynamic geometry software at the end of primary school, to work around properties of plane figures, such as invariant by dragging.

4.2.1. *A co-constructed coding*

This first example extends an exploration (with pupils aged 4–8 years), in the tangible space, of specific properties such as axial symmetry, convexity, curved or straight edges, the number of vertices or edges. These properties associated with the tangible space are characteristics whose formulation remains non-geometric.

According to the Plan d'Etude Romand (CIIP 2010), during the transition to Grade 3 (8–9-year-old pupils) geometric properties are introduced together with a conventional geometric lexicon. Non-conventional shapes, proposed by Vendeira-Maréchal and Coutat (2017), remain of interest. The properties that were initially made explicit in the tangible space using a non-geometric lexicon are made explicit in the graphical space using non-conventional signs. Here, the objective is to accompany the construction of geometric concepts with priority given to meaning. Therefore, the development of a shared coding between the pupils can participate in the abstraction of the characteristics of the shape in favor of the properties of the figure. However, there is no coding for the properties invested by the shapes in the collection (axial symmetry, convexity, type of edges and number of vertices). This coding is therefore developed in collaboration with the pupils.

The need for shared coding is motivated by a shape description communication task. Pupils and the teacher interact to collectively produce an economical way to make explicit different properties, regardless of the shape in which they appear. This work has led to the development of a set of cards. Figure 4.1 shows an example of describing a shape using specific cards. Each card color represents a property: the property of the edges of the shape (green card), the axial symmetry of the shape (yellow card), convexity (purple card) and the number of vertices (blue card). Illustrative drawings are used to identify the underlying property. These drawings can be used to build a lexicon and encourage the use of real or mental images. The association of a card for each color makes it possible to describe a set of figure properties.

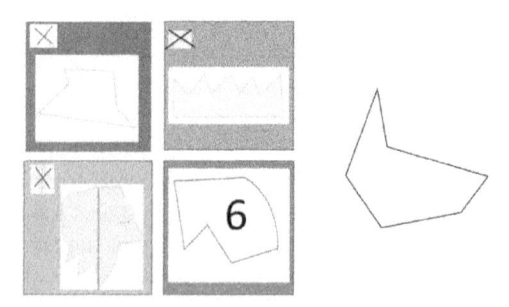

Figure 4.1. *Example of a figure corresponding to a card association. For a color version of this figure, see www.iste.co.uk/guille/tangible.zip*

These cards use the graphical register to represent properties which are mainly exploited via this register for 5–6-year-old pupils. However, the constraint of drawing, which freezes the objects represented, requires internal processing (in the sense of Duval [1993]) by the pupils to associate different representations of the same property. We could consider replacing the drawing with the explanation of the property in the discursive register (words), which would imply a conversion between the registers by the pupils. Cards using only a formal lexicon could be the next step.

In the next activity (Figure 4.2), pupils access a set of property cards and must find the figure corresponding to those properties from a set of four drawings. Pupils should not focus on the representation on the cards, but should be able to associate them with the representations on the cards presented. In the example presented, they must find a drawing that has straight and curved edges, has no axial symmetry, is non-convex and has five vertices.

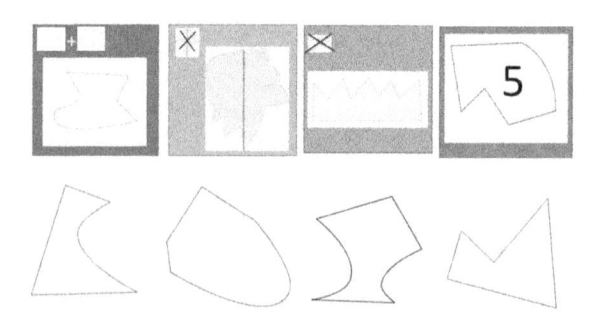

Figure 4.2. *Explanation of properties using shared coding. For a color version of this figure, see www.iste.co.uk/guille/tangible.zip*

The properties that are worked on here do not require the use of measuring instruments such as set squares or rulers. This means that several drawings can be associated with a collection of property cards. It is then a question of identifying contender drawings for a figure described by its properties.

These property cards can also complement freehand drawings, which are nevertheless too approximate to ensure the legibility of certain properties. They make it possible to bypass language or lexicon difficulties, and to consider the drawing as a possible (but not necessarily unique) representation of a figure.

4.2.2. Personal coding

The second example comes from a situation used for older pupils (10–12 years old) in a context that does not use manipulable shapes but uses dynamic geometry software (CabriElem®). Details of this research are presented in Coutat (2018). In this work, the use of CabriElem® allows working with properties as invariants of constructions: the properties of the geometric figure are invariant during the deformation of the Cabri-drawing (Laborde and Capponi 1994). Pupils worked with the construction and reproduction activities software, in order to appropriate the construction tools of the environment and, in particular, the dragging. Dragging allows access to properties associated with the use of specific construction tools. In the paper-and-pencil environment, the use of construction tools is indicated by coding. This is how coding was introduced to pupils: it allows access to the figure's properties through the use of construction tools. The coding of perpendicular and parallel lines as well as equal lengths was introduced.

In this example, pupils access a Cabri-drawing (Figure 4.3, left-hand side) and need to produce a message so that this Cabri-drawing is recognized among different drawings in the paper-pencil register. The drawing on the right also contains the coding for the properties of the figure.

Figure 4.3. *Cabri-drawing model*

Figure 4.4 shows the messages from three pupils. The coding of perpendicular lines (although worked on with the pupils), is not used. Pupils draw different configurations of the figure to illustrate the invariant properties of the figure.

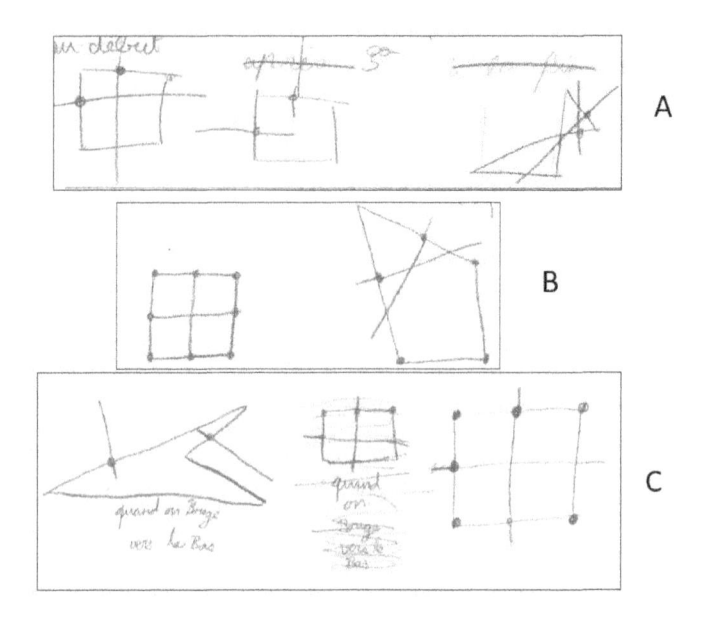

Figure 4.4. *Examples of pupil messages*

Unlike the previous example, conventional coding that completes a drawing is possible and was introduced to the pupils. The main difficulty is to statically code a dynamic invariance. The pupils do not use the coding of perpendicular lines even though the orientation of the lines drawn respects this perpendicularity. We can question the recognition of this property during the deformations of the Cabri-drawing. Here the pupils use a succession of drawings illustrating a particular configuration of the figure. The interest in the coding which allows for the reunion of all the representations of a property (here, perpendicularity), was not used by the pupils.

4.3. Conclusion: perspectives on the introduction of coding in geometry

In this chapter, we question the position of coding in primary school as a representation of geometric properties.

Coding can accompany a drawing to specify what is in the figure (see section 4.2.2. [Example 2]), or it can be used alone to communicate a property in the graphical register (see section 4.2.1. [Example 1]). In the case where the properties mentioned do not have a conventional coding, new signs can be constructed with the pupils (see Example 1). The introduction of coding in primary school is used here to distinguish, in the graphical register, between that which belongs to the figure (theoretical geometry) and that which belongs to observation ("tangible" geometry). Indeed, the graphical register is at the intersection of the two geometries and corresponds to the register used in the instrumented geometry practiced in the second half of primary school. In Example 1, the coding is constructed with the pupils. Each card defines a set of geometric figures with the attribute of the designated property. This communication medium is central to activities and cannot be bypassed by the pupils. In Example 2, pupils use a succession of drawings that are faithful to what is represented on their screen. It is difficult to know whether they identified the perpendicular lines without using the coding, or whether the visibility of the perpendicular lines on their drawings is only the result of the precision of the pupils' tracing abilities. In both examples, the coding makes it possible to encompass all the possible representations of the property into a single representation. There is then an internal transformation exercise in the graphical register that consists of allowing different configurations for a property. In the activity illustrated in Figure 4.2, this translates into an analysis of the drawings to detect the presence or absence of the properties explained by the cards. In Example 2, this translates into the possible static representation of an invariant relationship between geometric objects during dragging.

The use of coding allows us to understand the figure as detached from what the drawing shows (Duval 1994). This evolution in the understanding of the figure, through exploration in a dynamic environment, may be an explanation for the non-use of the coding of perpendicular lines in the pupils' drawings. The use of coding for the introduction of geometric properties in primary school accompanies the change in view of geometric objects. The internal treatment of the graphical register of representations participates in the evolution for understanding the figure. It would be interesting to extend this study to introduce a formal lexicon to progressively replace coding, and to study the impact of a conversion register work (graphical/discursive) on the pupils' geometric knowledge.

4.4. References

CIIP (2010). Plan d'étude romand [Online]. Available at: https://www.plandetudes.ch/ [Accessed 16 September 2021].

Coutat, S. (2018). Expression des propriétés géométriques entre géométrie statique et géométrie dynamique. *Actes du 44e Colloque de la COPIRELEM*, ARPEME, Paris.

Duval, R. (1993). Registres de représentation sémiotique et fonctionnement cognitif de la pensée. *Annales de didactique et de sciences cognitives*, 5, 37–65.

Duval, R. (1994). Les différents fonctionnements d'une figure dans un raisonnement géométrique. *Repère IREM*, 17, 121–138.

Houdement, C. and Kuzniak, A. (1998). Réflexion sur l'enseignement de la géométrie pour la formation des maîtres. *Grand N*, 64, 65–78.

Houdement, C. and Petitfour, E. (2018). Malentendus sémiotiques dans l'enseignement spécialisé. *Actes du 44e Colloque de la COPIRELEM*, ARPEME, Paris.

Laborde, C. (2004). Come la geometria dinamica puo rinnovare i processi di mediazone delle conoscenze matematiche nella scuola primaria. In *La didattica della matematica: Una scienza per la scuola*, D'Amore, B. and Sbaragli, S. (eds). Pitagora, Bologna.

Laborde, C. and Capponi, B. (1994). Cabri-géomètre constituant d'un milieu pour l'apprentissage de la notion de figure géométrique. *Recherches en didactique des mathématiques*, 14(1–2), 165–210.

Salin, M.-H. (2003). Comprendre les difficultés des élèves à passer de la "géométrie de l'école primaire" à la "géométrie du collège" [Online]. Available at: http://emf.unige.ch/files/9614/5459/5580/EMF2003_GT5_Salin.pdf.

Vendeira-Maréchal, C. and Coutat, S. (2017). "C'est une montagne ou une trompette ?" Entre perception globale et caractéristiques des formes aux cycles 1 et 2. *Grand N*, 100, 79–104.

Vergnaud, G. (1990). La théorie des champs conceptuels. *Recherches en didactique des mathématiques*, 10, 133–170.

Verin, A. (1995). Mettre par écrit ses idées pour les faire évoluer en sciences. *Repères*, 12, 21–36.

5

Freehand Drawing for Geometric Learning in Primary School

5.1. Introduction

This research extends the work of Vendeira-Maréchal and Coutat (2017) on the preliminary learning of geometry in primary school. Noting that work on the characteristics of shapes is possible from the age of 4, the next question is what to do next for pupils' geometric learning at age 8. The resource developed in the framework of our work for learning in early education included exclusively activities focused on the recognition and classification of geometric shapes. We are therefore developing tasks involving the construction and reproduction of geometric figures for pupils 8 to 10 years old. However, in Geneva, new teaching methods will come into force in 2022, in classes for pupils 8 to 10 years old (CIIP 2018). Geometry instruments will make their first appearance there. Until then, they only appeared at the age of 11, in construction tasks that require knowing how to construct/complete polygon properties and also the way in which to use geometry instruments correctly. The construction of parallel and perpendicular lines was thus worked on in the classrooms via steps to follow (an example of a sheet in the pupil's notebook is proposed in Figure 5.1).

In the new teaching resources for 8- to 10-year-olds, emphasis is also placed on the use of geometric instruments to carry out constructions with precision, and also to verify and question intuitions. For 8- to 9-year-olds, the notion of right angles is worked on through the introduction of the set square, and the notion of parallel lines is added for 9- to 10-year-olds. It is therefore in this context of Geneva schooling, when the use of geometric instruments will be introduced to younger pupils, that we wish to investigate the potential of freehand drawing. Our research aims at

Chapter written by Céline VENDEIRA-MARÉCHAL.

questioning its impact and relevance in construction/reproduction tasks with 8- to 10-year-olds, in particular to free them from the manipulative constraints implied by geometry instruments.

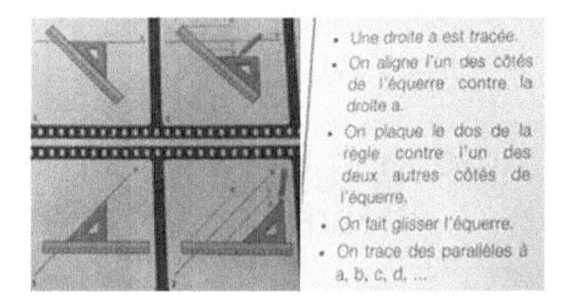

Figure 5.1. *Extract from a notebook on Swiss francophone teaching resources*

5.2. Drawings in geometry and their functions

In primary school, during construction/reproduction tasks of geometric figures, drawings of different natures coexist. As Laborde and Capponi (1994) point out, *drawing* is a representation of a *figure* through a material trace (whether on a sheet of paper or a screen). In our research, we distinguish between two main categories of drawing in geometry: instrument drawing and freehand drawing. The instrumented drawing is made with drawing instruments (the ruler and set square in particular, and also templates and grids), which are themselves carriers of properties. Drawings produced in a computer environment also fall into this category. However, freehand drawing does not involve the use of instruments. It also does not require coding or accompanying text to highlight its properties, although these may sometimes accompany it.

Chaachoua (1997) distinguishes, in geometry, at the secondary level, three levels of drawing intervention (instrumented or freehand), to which he associates specific functions:

– intervention at the level of the statement to take charge of hypotheses, and/or to make visible the figure or a subfigure relevant to the resolution;

– intervention in the solution to illustrate, and/or to explore the situation, to give a counterexample, and to stimulate conjecture;

– intervention in the solution, in order to illustrate the steps, to leave visible the construction lines as traces of the procedure used, and to answer the problem.

In primary school, the three levels of intervention are different from the secondary school context. However, we have adapted them to our context and focused on freehand drawing. With 8- to 10-year-old pupils, when a drawing is present on an exercise or accompanies an instruction, it is rarely done freehand. We encounter more this case in secondary school, where the underlying intention is to force the pupil to put aside their visual perception and thus to consider the ideal object, whatever its representation (Coppé et al. 2005). In our research, the drawing is used at the level of the statement as a model to be reproduced and is not, therefore, done freehand, so that the pupils can extract the information (properties) so as to reproduce it. During the resolution phase, primary school pupils rarely need to produce drawings in order to represent the problem posed. It is therefore not the assistance function that interests us, but the experimentation function. In our research, freehand drawings are used as a field of experimentation: as the pupils are freed from material constraints, we hypothesize that they can thus produce drawings more freely and potentially more quickly. Finally, it does not seem relevant to us to consider freehand drawing as the solution to a geometry problem because it does not intrinsically carry any property nor does it offer the possibility to grasp the construction process. However, the succession of intermediate productions allows us to reconstruct the stages that led to the final product, and to access the procedures implemented. It is thus the different stages of construction/reproduction of the figure that interest us rather than the finished product. In our research, freehand drawings have two distinct functions: to experiment without manipulative constraints; and to produce, as much as is necessary, in order to regulate/transform one's knowledge into a construction.

5.3. Freehand drawing in research

There is little French-language research on the teaching learning of geometry involving freehand drawing. Coppé et al. (2005) were interested in the different types of drawings in argumentation activities in the 5th grade: they defined five types of drawings, including freehand drawings. However, freehand drawing is essentially studied through its use in school textbooks. It would help to promote the transition between practical and theoretical geometry. Conne (2006) studies freehand drawings produced by special education pupils: they have to produce cubes or other polyhedra from a model or from memory. Conne observes that the problems encountered require pupils to question "their models, drawings and procedures" (op. cit., p. 252, *translated by author*). He adds that freehand drawings, which are easy to make, are immediately seen as something that can be repeated several times. The freehand drawing is then considered in a dynamic process where the knowledge is built and transformed during the course of the productions. Several works focusing on freehand drawing production exist in the anglophone literature (Brooks 2009; Sinclair and Gol Tabaghi 2010; Thom and McGarvey 2015; Dalh 2019; Way and

Thom 2019). Thom and McGarvey (2015) analyze the work of pupils, aged 4–7 years old, in three geometry situations involving freehand drawing. They ask whether interpreting the drawings could help them understand children's geometric thinking, and thus consider drawings as a vehicle for thinking and not just as an object of reasoning. Drawings could thus be used by researchers as artifacts to highlight what children "know", including their cognitive abilities, spatial awareness and geometric knowledge (op. cit., p. 467, *translated by author*). They also hypothesize that the reproduction tasks would allow pupils to conceptualize geometric objects. The analysis of the three geometric situations proposed in this research illustrates that the children's geometric reasoning unfolds "in the flow of the drawing" (op. cit., p. 479). This supports the idea that freehand drawing is seen as a dynamic process. As Conne (2006) also mentions, "in his sketches, the draftsman alternately leaves the drawn object for the process and then returns to his model. He thus builds on the qualities and defects of his sketches as well as his conceptions" (op. cit., p. 253, *translated by author*). A focus sometimes on the control of freehand tracings and sometimes on the properties that define the figure to be reproduced leads us to draw a parallel with the work developed by Berthelot and Salin (1994) around spatial and geometric knowledge.

In view of these few contributions, we hypothesize that in freehand reproduction tasks in geometry, the drawings produced in a dynamic process inform us about the pupils' spatial and geometric knowledge as well as their capacity to mobilize them together.

5.4. Exploring the milieu around a freehand reproduction task of the Mitsubishi symbol on a blank white page

Our classroom experiments are based on a set of tasks (Favre 2008) around the Mitsubishi symbol. It was developed for pupils in both special education and regular education in the cantons of Geneva and Vaud in Switzerland. In what follows, we develop only one of the tasks of the set, namely, "to find several ways of drawing freehand the Mitsubishi symbol (Figure 5.2) on a blank white page with the model available". An exploration of the milieu (Conne et al. 2003) around the Mitsubishi symbol reveals, among other things, hidden mathematical knowledge, and the different tasks and hints that can be envisaged for pupils. For example, in order to encourage pupils to find several ways of reproducing the symbol, we ask them to find out how to reproduce it as quickly as possible, in one stroke of the pen, in the most precise way, etc. Only when they have done so will they be able to see the symbol. Challenging the pupils allows us to enrich the set of tasks by stabilizing certain tasks and revealing new paths. The process and the pupils' productions also contribute to the exploration of the milieu. The description below of the different ways of drawing the Mitsubishi symbol freehand, on a blank white page, with the model, constitutes the current state for the exploration of the milieu of this task.

Figure 5.2. *Symbol to be reproduced freehand on a blank sheet of paper*

A possible procedure using freehand drawing consists of considering three independent rhombuses (or parallelograms) (Figure 5.3A). In this case, it is the "surfaces" vision (Duval and Godin 2005) that is predominant, although its realization implies a 1–2D articulation: it is a question of realizing rhombuses (or parallelograms) and of thinking about their arrangement in relation to each other, involving parallelism controls, segment extensions, alignments, etc.

There is also the question of the order in which the symbol will be reproduced: for example, we can assume that the rhombus, on its point, will be the first because it is in its prototypical position, and its axes of symmetry are arranged horizontally and vertically in relation to the edges of the sheet. The other two rhombuses are not in their prototypical position, so it is likely that the task becomes more complex when it comes to reproducing these. These few developments show that drawing from a set of possibilities is not enough to understand what is really going on. This allows us, however, to realize a set of relations which could be at work when a pupil is instructed to draw freehand. Without going into too much detail, as in the previous example, it is also possible to consider networks of intersecting lines and thus adopt a "line" vision. To create the symbol without lifting the pencil, it is necessary, for example, to act on a network of lines by checking the intersections, the segment extensions of the aligned sides and the parallelism (Figure 5.3B). It is also possible to start with three intersecting segments to which "hats" are added (Figure 5.3C). In this case, the three rhombuses are connected by their vertex.

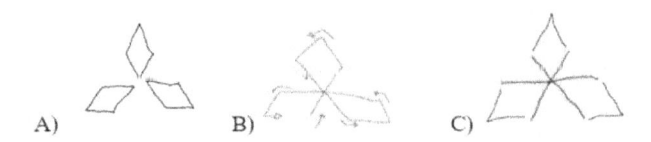

Figure 5.3. *Three ways of reproducing the Mitsubishi symbol: according to three distinct and independent parts (A), according to a network of straight lines, (B) without lifting the pencil or (C) from three intersecting segments*

Another procedure consists of starting with an equilateral triangle, from which three pieces are removed (Figure 5.4A): the "surfaces" view takes priority with a decomposition of one surface into several others. In the following example (Figure 5.4B), we start with a large equilateral triangle from which we remove only one part, thus creating a "teepee"; next, we draw straight lines to obtain the symbol.

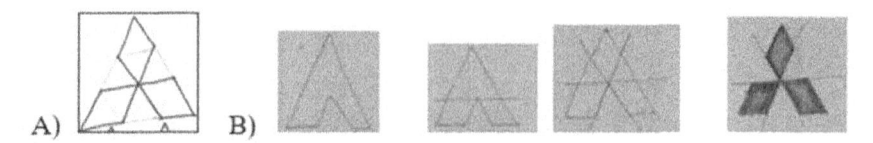

Figure 5.4. *Two ways to by which reproduce the Mitsubishi symbol from a triangle*

Another way to reproduce the symbol is to reconstruct a triangular network to show the symbol. In this case, the network produced freehand allows a shift to the instrumented drawing whose embedded properties have been constructed by the pupil. Several methods are possible. We can construct a triangular network within the basic triangle (Figure 5.5), which is like paving an equilateral triangle into smaller units. The procedure involves dividing the sides of the starting triangle into equal parts, and then drawing lines parallel to the sides of the existing triangle. The more you divide the sides of the starting triangle, the more consistent the pavement is.

Figure 5.5. *Reproduction of the symbol by creating a triangular network inside an equilateral triangle*

It is also possible to create a triangular network from a double network of parallel lines (Figure 5.6A) or from a triangle to which a new equilateral triangle is attached on each of its three sides and so on (Figure 5.6B). Without being able to anticipate the different procedures that may emerge from pupils confronted with this task, it seems that the mere fact of experimenting with several of them leads them to gradually take control of the milieu by engaging in geometrical reasoning. In what follows, we narrate two experiments (Favre 2015, 2019), with two different pupils faced with this task.

A)   B)

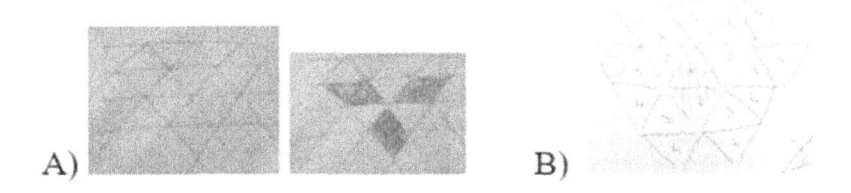

Figure 5.6. *Two methods of reproduction by creating a triangular network*

5.4.1. *Freehand drawing reveals a reasoning between spatial knowledge and geometric knowledge*

This first example highlights the different geometrical reasonings when a pupil uses freehand drawing in a problem-solving situation. How can we consider the drawings produced? What does they tell us about the pupils?

We propose a possible interpretation of a pupil's reasoning produced when doing the task described above. From the production (Figure 5.7), we can hypothesize that the pupil reproduces the symbol by considering its three parts separately, as three independent rhombuses (or parallelograms). According to our hypotheses, he first constructs a first rhombus, the one in its prototypical position. Then, for the right-hand rhombus, the pupil extends one of the sides of the first rhombus, taking into account their alignment. In this second reproduced rhombus, the parallelism of the sides is also respected. However, the length of the sides is no longer controlled by the length of the first rhombus. The figure is still a rhombus. The third rhombus is likely completed using the same procedures with respect to the parallelism of the sides and certain alignments. However, there are more alignments to be made in this last part which could justify that they are not all considered. The position of the last reproduced part is controlled, but neither the size nor the shape of the rhombus is respected. From the first two rhombuses created, the pupil has no choice but to respect the alignments and parallelism and produce a last rhombus that visually approximates the symbol without taking into consideration all the geometric relations (Figure 5.7).

Two types of control seem to be necessary in order to carry out this task: knowledge and control of several properties (equality, parallelism, alignment, measurement, etc.), representative of geometric knowledge, and the perceptual control "it looks like", which is more representative of spatial knowledge. When there is no coordination between these two types of control, it is possible to deviate significantly from the global form of the Mitsubishi symbol, which is not the case in this example.

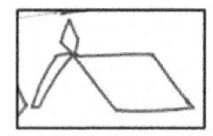

Figure 5.7. *Freehand production of the Mitsubishi symbol by a pupil*

5.4.2. *Freehand drawing as a dynamic process to build and transform knowledge*

In the following example, we highlight the way in which the pupil's reasoning evolves during the lesson following the successive productions of freehand drawings. The pupil seems to gradually take control of the milieu, in coordinating the properties with the global vision of the model to be reproduced. The order in which the pupil carries out their drawing seems to be determined by the properties of the geometric object they have to reproduce (the symbol), and also by the constraints of the triangular network (Duval 1994). As in the previous example, this is a narrative involving some interpretation. The pupil locates the first rhombus, the top one, in its prototypical position. He then completes the picture with two rhombuses located below. He uses the vision of three independent equal rhombuses. He produces a first drawing (Figure 5.8) by coloring the outline of the reproduced figure. He does not check the orientation of the other two rhombus. He identifies the figure according to intrafigurational indicators: "difference in size or orientation between certain figurative units constituting the image of what is seen" (ibid., p. 124, *translated by author*). But were the gestures not controlled? There is, indeed, the presence of three distinct parts, the alignment between the sides of the various rhombuses and the respect to opposing parallel sides within the three parts of the symbol. This control is facilitated by the presence of the network. The network therefore has as much a facilitating effect as it does a constraining effect on the task. The pupil lacks the control for the orientation of the two lower rhombuses, hence the consideration of an alignment of these along a horizontal axis.

Figure 5.8. *First freehand production by a pupil*

The question, "Is this the Mitsubishi symbol?" leads the pupil to consider the produced drawing in a global way. This global view sanctions the production and leads the pupil to a new drawing (Figure 5.9A). In this second production, we see that the alignment along the horizontal axis that allows for the reproduction of one of the four sides of the bottom rhombuses is well respected. The pupil also checks that this is a non-convex figure (which was partially missing in his first production).

Here, we see an intersection from which three "diamonds" radiate. It is probably to take into account the orientation of the two lower parts that he produces two triangles instead of rhombuses. When asked "Is this (the top rhombus) the same as this (one of the triangles produced)?" the pupil invalidates his production by using geometric arguments about the nature of the shapes produced: "Because it's not a diamond". He then corrects his erroneous production by transforming the triangles into rhombuses. He then finalizes the symbol by fully coloring it in (Figure 5.9B).

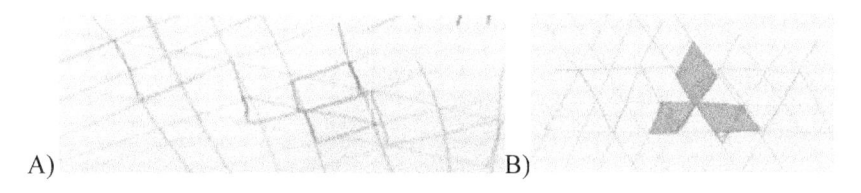

Figure 5.9. *Second pupil production (A) freehand and then (B) corrected*

5.5. Conclusion

The two examples presented allow us to infer that freehand drawing in geometric problem solving is not restricted to the function of illustration. It allows the development (for the pupil) and the understanding (for the researchers) of a form of geometric thinking. This geometric thinking is characterized by geometric and spatial knowledge (Berthelot and Salin 1994). In the presented task, the geometric knowledge concerns the elaboration and use of geometric concepts (alignment, parallelism, length of segments, angles) and spatial knowledge control the production of freehand drawings. However, our experiments show that appropriate reasoning does not guarantee the reproduction of a valid freehand drawing. Indeed, in the action, something more complex is at stake since the pupil has to coordinate all the elements on which he can reason. Therefore, geometric and spatial knowledge must be coordinated. Our research shows that this process can be facilitated through the production of a series of freehand drawings, which would allow the pupil to gradually take control of the milieu by effectively coordinating spatial and geometric knowledge.

The freehand drawings are quick to execute and the pupils can produce many of them, in order to regulate/transform knowledge by a game of successive trials and adjustments. Thus, all the traces collected testify to the knowledge implemented in a back and forth between the drawings produced and the desired drawing. Consequently, the drawings allow us to learn about the state of knowledge of the properties and spatial relations of the students, and are therefore indeed "vehicles of thought".

The freehand drawings are quick to execute. Pupils can produce many of them, in order to regulate/transform their knowledge through successive trials and errors. Therefore, the set of drawings collected testifies the knowledge implemented between drawings produced and the desired drawing. Consequently, the drawings allow us to learn about the state of the pupils' knowledge of properties and spatial relations, and are therefore "vehicles for thought".

5.6. References

Berthelot, R. and Salin, M.-H. (1994). L'enseignement de la géométrie à l'école primaire. *Grand N*, 53, 39–53.

Brooks, M. (2009). Drawing, visualisation and young children's exploration of "big ideas". *International Journal of Science Education*, 31(3), 319–341.

Chaachoua, H. (1997). Fonctions du dessin dans l'enseignement de la géométrie dans l'espace. Etude d'un cas : la vie des problèmes de construction et rapports des enseignants à ces problèmes. PhD Thesis, Université de Grenoble 1, Grenoble.

CIIP (2018). ESPER CIIP [Online]. Available at: https://www.ciip-esper.ch/.

Conne, F. (2006). Quelques pas esquissés dans l'univers des polyèdres : suivi de situations de dessin en 4ème primaire de l'enseignement spécialisé. In *Difficultés d'enseignement et d'apprentissage des mathématiques. Hommage à Gisèle Lemoyne*, Giroux, J. and Gauthier, D. (eds). Editions Bande Didactique, Montréal.

Conne, F., Cange, C., Favre, J.-M., Del Notaro, L., Scheibler, A., Tièche-Christinat, C., Bloch, I., Salin, M.-H. (2003). L'enseignement spécialisé, un autre terrain de confrontation des théories didactiques à la contingence. In *Actes du séminaire ARDM de didactique des mathématiques*, Durand-Guerrier, V. and Tisseron, C. (eds). IREM Paris 7, Paris.

Coppé, S., Dorier, J.-L., Moreau, V. (2005). Différents types de dessins dans les activités d'argumentation en classe de 5ème. *Petit x*, 68, 8–37.

Dahl, H. (2019). "He's so fast at drawing" – Children's use of drawings as a tool to solve word problems in multiplication and division. *Eleventh Congress of the European Society for Research in Mathematics Education*, Utrecht University, Utrecht.

Duval, R. (1994). Les différents fonctionnements d'une figure dans un raisonnement géométrique. *Repère IREM*, 17, 121–138.

Duval, R. and Godin, M. (2005). Les changements de regards nécessaires sur les figures. *Grand N*, 76, 7–27.

Favre, J.-M. (2008). Jeu de tâches : un mode d'interactions pour favoriser les explorations et les expériences mathématiques dans l'enseignement spécialisé. *Grand N*, 82, 9–30.

Favre, J.-M. (2015). Investissements de savoirs et interactions de connaissances dans un centre de formation professionnelle et sociale : une contribution à l'étude des mathématiques dans le contexte de la formation professionnelle spécialisée. PhD Thesis, Université de Genève, Geneva.

Favre, J.-M. (2019). Expérience et interprétation – Faire des mathématiques avec des élèves de l'enseignement spécialisé. *Actes des troisièmes journées didactiques de La Chaux-d'Abel 3-4-5 mai 2018.*

Laborde, C. and Capponi, B. (1994). Cabri-géomètre constituant d'un milieu pour l'apprentissage de la notion de figure géométrique. *Recherches en didactique des mathématiques*, 14(1.2), 165–210.

Sinclair, N. and Gol Tabaghi, S. (2010). Drawing space: Mathematicians' kinetic conceptions of eigenvectors. *Educational Studies in Mathematics*, 74(3), 223–240.

Thom, J.S. and McGarvey, L.M. (2015). The act and artifact of drawing(s): Observing geometric thinking with, in, and through children's drawings. *ZDM Mathematics Education*, 47, 465–481.

Vendeira-Maréchal, C. and Coutat, S. (2017). C'est une montagne ou une trompette ? Entre perception globale et caractéristiques des formes aux cycles 1 et 2. *Grand N*, 100, 79–104.

Way, J. and Thom, J.S. (2019). Capturing the mathematical drawing process using a digital pen. *Annual Meeting of the Mathematics Education Research Group of Australasia*, Perth.

PART 2

Resources and Artifacts for Teaching

6

Use of a Dynamic Geometry Environment to Work on the Relationships Between Three Spaces (Tangible, Graphical and Geometrical)

This chapter has three main objectives: to show some results from my research on the uses of a Dynamic Geometry Environment (DGE), which are old but still relevant, to expose a set of situations for first graders (6 years old; first year of primary school) and to relate these results to more recent research on the uses of this type of software to work on the relationships between tangible space, graphical space and geometrical space.

6.1. Added value with a dynamic geometry environment: the ecological and economical point of view

Since the end of the 1980s, a lot of research has been done on the contribution of Dynamic Geometry Environment for the teaching of geometry at different levels, from primary school through to university.

Soury-Lavergne synthesizes some of this work in two recent publications: one with Maschietto, in a course of the 20th Summer School for Teaching Mathematics (Soury-Lavergne and Maschietto 2017), the other in a report on Dynamic Geometry for teaching and learning mathematics at the request of the Centre National d'Étude du Système Scolaire (CNESCO) (Soury-Lavergne 2020). In this last publication, Soury-Lavergne (2020) indicates that teachers do not make sufficient use of dragging, which is nevertheless one of the most valuable features of this type of

Chapter written by Teresa ASSUDE.

software, as research shows. This result is all the more questionable since the work on the importance of displacement is not recent. The dissemination of these results to teachers is a problem that educators are asking themselves (Lagrange and Caliskan-Dedeoglu 2009). Personally, I have disseminated some of these works in the context of initial or in-service training of primary school teachers, but obstacles and resistance from teachers have not always allowed this type of software to be properly used by a large part of these professionals in training: relationship to digital technology, management of a class using digital technology, maintenance and technical problems and also the relationship to geometry of some primary school teachers.

This chapter begins with the presentation of my work on the use of a Dynamic Geometry Environment in primary school classes in France. The point of view adopted is that of the economy and ecology of mathematical work. Indeed, when a tool is introduced in the framework of a work system (Chevallard 1992), it must be considered with the other tools so that it is ecologically viable within the teaching system, with the constraints and the possibilities of that system. The viability for the introduction of a new artifact is also linked to the added value that it brings in relation to other tools, including but not limited to the added teaching value (Assude and Loisy 2009). In the case of Dynamic Geometry, the added value is based on several elements, as we have presented (Assude et al. 1996): the possibility of drawing a large quantity of drawings, the speed of execution, which makes possible graphical experiments and the observation of invariants amongst these drawings by moving the points to find figures (Laborde and Capponi 1994; Hölz 1996; Arzarello et al. 2002). A figure, here, is a class of invariant drawings, according to determined properties (Laborde and Capponi 1994). This added value is also that DGE software allows the creation of new tools (the macro-constructions of Cabri-geometry, for example) and new situations for teaching, such as "soft locus" (Assude et al. 1996).

These new situations that distinguish between soft and hard constructions have been developed by several researchers (Hölz 1996; Healy 2000; Laborde 2005), as indicated by Soury-Lavergne (2020). These new situations (e.g. black boxes in the sense of Laborde and Capponi (1994)), open up possible avenues for pupils' mathematical learning that were hardly exploited or exploitable with other tools. The old-new dialectic is thus essential to make working with this type of software viable (Assude and Gélis 2002). This dialectic can be thought of in several ways: old types of tasks with old and new tools, which allows for different techniques (old and new), and new types of tasks with old and new tools, which also allows for new techniques, even though the tools are old. The old-new articulation is thus a way to make mathematical praxeologies (Chevallard 1999) and teaching situations evolve. This is indeed the case with the "dragging" tool in a DGE software. This tool makes it possible to explore situations, to validate and invalidate constructions, and to identify invariants, among others (Arzarello et al. 2002; Goussseau-Coutat 2006; Restrepo 2008). We have studied this old-new dialectic in a teaching sequence on quadrilaterals with classes CM1 and CM2 (pupils aged 9 and 10) (Assude and Gélis 2002). The symbiosis between the different tools and the intertwining of praxeologies allows the pupil to encounter different types of tasks and techniques (old and new) using several tools. One example studied was the construction program using pencil and paper, and the history that existed in the Cabri version at the time. This work allowed pupils to encounter and construct quadrilaterals, and also to establish relationships between particular quadrilaterals: for example, pupils would open a Cabri file where they saw a square that was constructed as a rhombus (or rectangle). By moving points around, pupils realized that the drawing was not a square, but that the figure was a rhombus. Therefore, they saw the square as a particular rhombus when the points were in a certain position, where characteristic properties of the square were verified. This type of work was also indicated by Sinclair and Bruce (2015) as potentially having an important role for pupils to work on the properties of figures:

> DGE affordances suggests that they could both (1) help learners see and make a large example space of geometric shapes such as long, skinny triangles, and (2) help learners appreciate aspects of the inclusive relations in the sense that it is possible to transform a constructed parallelogram, for example, into a rectangle (Sinclair and Bruce 2015, p. 325).

This work on quadrilaterals allowed us to identify a condition for the viability of the integration of a DGE in the mathematical work of the pupils, which is the "right distance" between the old practices (or praxeologies) and the new practices (praxeologies). By analyzing the work of two teachers over a period of two years, we were able to observe that they did not completely change their practices – as they said during the interviews – but that they adjusted the old practices to this new tool: the DGE. It seems necessary to us to take into account this ecological condition in the framework of teacher training, and in the framework of the design and use of resources: to start from existing practices and to see how to make them evolve through this old/new dialectic, and the symbiosis between the different tools as well as the praxeological interweaving.

Another condition for the new to have its place in symbiosis with the old is that of time economy (Assude 2005). Chevallard and Mercier (1987) have shown the strong constraint of didactic time, that is, the time of the advancement of knowledge. The management of didactic time is a professional problem, and the introduction of a new tool provokes changes to be taken into account within the time economy of a teaching system. In the 2005 article, we showed a number of time-saving strategies that allow the teacher to make the symbiosis between the different tools of the pupil's mathematical work and to save time. The time-saving strategies we identified are as follows:

– Manage didactic time in a way that saves time by getting to the heart of what needs to be known.

– Alternate the types of tasks proposed to the pupils: construction tasks, figure analysis tasks such as those in Figure 6.1. This example from the sequence on quadrilaterals shows that the pupils analyze the initial figure, move the points, observe the transformations and identify relationships (properties and inclusion of particular quadrilaterals). By having the pupils analyze the figures, the teacher saves time.

– Alternate phases of individual and group work.

– Fix the movement during the exploration of the figures, by setting up ostensives (e.g. a table, Figure 6.2) that condense the essential information that the pupil must know. In the same situation involving quadrilaterals, after exploring and analyzing the figures, the teacher proposes a group phase in which the pupils identify the essential information and determine what they need to know.

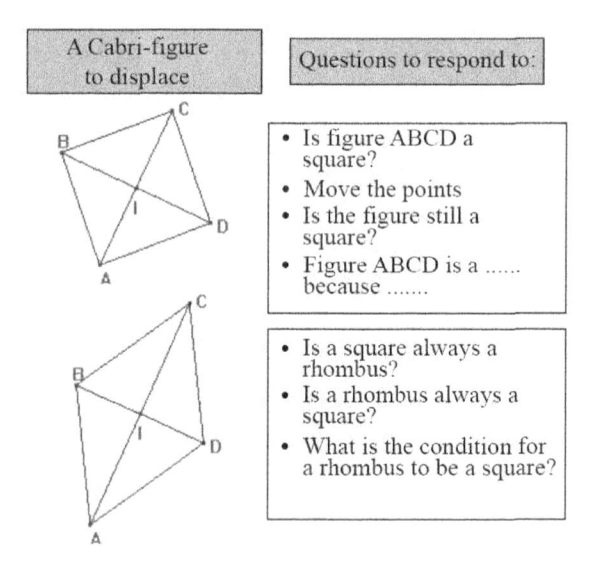

A Cabri-figure to displace	Questions to respond to:
(figure of quadrilateral ABCD)	• Is figure ABCD a square? • Move the points • Is the figure still a square? • Figure ABCD is a ...... because ....
(figure of quadrilateral ABCD)	• Is a square always a rhombus? • Is a rhombus always a square? • What is the condition for a rhombus to be a square?

Figure 6.1. *Example of an analysis task*

The importance of time economy in the framework of Dynamic Geometry has been subsequently worked on by other researchers who take into account a temporal axis, such as Ruthven (2009), Abboud-Blanchard (2013) and Abboud-Blanchard and

Rogalski (2017). The latter, while using a different theoretical approach, recover some of these results related to temporal management, showing their robustness.

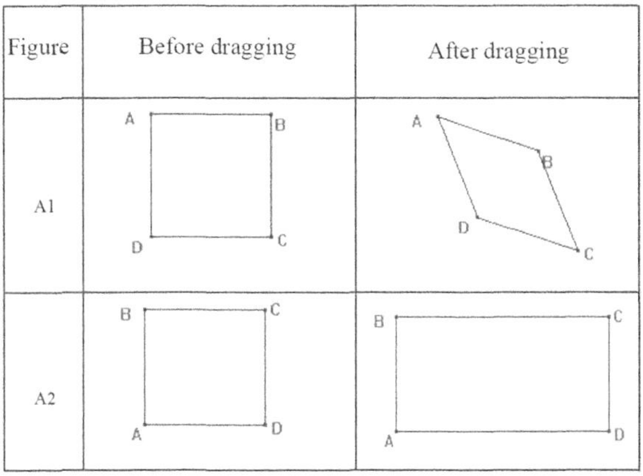

Figure	Before dragging	After dragging
A1	square ABCD	distorted quadrilateral ABCD
A2	rectangle BCDA	rectangle BCDA (wider)

Figure 6.2. *Ostensives*

These results are not necessarily known by teachers. Abboud-Blanchard and Rogalski (2017) and Soury-Lavergne (2020) show that teachers have an effective relationship with DGE, but do not use the potentialities of this artifact, in particular the movement of points or objects. One question that arises is how and to what extent this type of software is integrated into the classroom (Assude 2007). We designed an observation and analysis grid based on several indicators: praxeologies (types of tasks, techniques, discourse); tools (artifact/instrument); contract/environment; temporalities; and relationships between some of these indicators (Assude 2007). From these indicators, we have defined several modes of integration. We present here two of them: the instrumental integration mode and the praxeological integration mode.

The *instrumental integration mode* reflects the way in which the instrumental dimension is taken into account. This mode of integration can be broken down into four modes: the *instrumental initiation* mode, when the types of tasks are focused on initiation to the artifact (the keys, how it works, etc.); the *instrumental exploration* mode, when the types of tasks can be mathematical but the aim is to explore elements of the functioning of the artifact; the *instrumental reinforcement* mode, when the types of tasks are mathematical but a one-off contribution of information on the artifact is necessary, and the *instrumental symbiosis* mode, when the types of tasks are mathematical and the artifact is no longer a source of problems and is in symbiosis with the mathematical work.

The *praxeological integration mode* (Assude 2007) relates to the praxeological work proposed to pupils, to the relationship between praxeologies with/without a tool (in particular, DGE). This mode of integration can be broken down into three modalities: the *minimal praxeological* mode, when instrumental types of tasks and techniques are targeted (which may correspond to instrumental initiation); the *praxeological juxtaposition* mode, where the types of tasks and techniques with the software are juxtaposed with the types of mathematical tasks and paper-and-pencil techniques, without there being any relationship between them; and the *praxeological intertwining* mode, where the types of tasks and techniques of software and paper-pencil are intertwined, but where the techniques are not justified by a technologic-theoretical discourse (as per Chevallard). If the techniques are justified then the mode of praxeological integration is maximal. These modes of integration can be reconciled with the work of Soury-Lavergne and Maschietto (2019) and Voltolini (2017) on the "artifact duo". This work also defends the hypothesis that one of the conditions favorable to pupils' learning with DGE is that we should "seek an articulation with different geometric problem-solving environments, especially environments relying on tangible material" (Soury-Lavergne 2020, p. 19, *translated by author*).

These two modes are associated with an old/new dialectic and with a contract "milieu" dialectic. In fact, the introduction of new tools into the environment can allow, through their added value, an experimental and control element to work that can be complementary to the one done with typical tools. The possibility of a change in the teaching contract is thus an added value for this type of software (Assude and Gélis 2002), even though this is not always the case, as shown by the Soury-Lavergne (2020) survey.

6.2. Tangible space, graphical space and geometric space

Since the 1980s, research in mathematics teachings has focused on the relationship between tangible space and geometry, as well as on situations that take these relationships into account in primary and secondary school teaching methods.

The first of these works, those of Brousseau (1983) and Berthelot and Salin (1992), insists on the importance of three issues: practical, geometric and modeling. They set up teaching situations for pupils to experience these relationships. Generic models, such as geometric paradigms (Houdement and Kuzniak 2006) or geometric workspaces (Kuzniak 2007), have also made it possible to think about these relationships. We will not synthesize this research; for that see Sinclair and Bruce (2015), Mathé and Mithalal-Le Doze (2019), Houdement (2017), Soury-Lavergne and Maschietto (2019) and Soury-Lavergne (2020). However, we will present a very fragmented selection of research works that have been interested in the relationship between the three spaces: tangible space, graphical space and geometric space, in relation to the use of a DGE. The results of this work will then allow us to justify the principles that are at the basis for the design of the experimental situations that we present in section 6.3, and to allow us to ask ourselves a certain number of questions in the conclusion.

Houdement (2019) indicates that in France, the "spatial is studied from two angles: to equip the citizen in his relationship to the tangible space and to specify his relationship with the geometric" (p. 19, *translated by author*). According to this author, all the potentialities of this work are far from being exploited in teaching. For example, Gobert (2001) was interested in the links between spatial and geometric knowledge by proposing situations in micro- and meso-space. However, this type of situation is not often used in teaching, even though resources exist, such as those in the ERMEL collection (ERMEL 2006, 2020) or the work of Emprin (2014) on the simulation of displacements.

The Lille group has been interested for years in the problems of reproducing figures, tracings, the role of instruments and constraints (Perrin-Glorian et al. 2013; Perrin-Glorian and Godin 2014; Mathé and Mithalal-Le Doze 2019). The place of instruments (tools) is essential in the situations experimented with pupils, by the constraints they induce: these authors call this the "geometry of tracings", this geometry of geometric constructions where we are interested in the constraints imposed by the instruments and their usage cost. These instruments are not only the "usual" tools, such as the ruler and the compass, but also can be very different, as shown by the research carried out on templates (Coutat and Vendeira-Maréchal 2015).

Perrin-Glorian and Godin (2018) highlight the relationships between the three spaces, physical tangible space, geometric space and graphical space, the latter having a "pivotal role" between the other two. In the same sense, Chevallard and Jullien (1991), Mercier and Tonnelle (1992) or Bloch and Pressiat (2009) insist on the role of schematization and semiotic tools, as a means of modeling objects in tangible physical space. For these authors, schematization has a pivotal role in the passage between tangible physical space and geometric space, and should be the object of particular attention in primary school. These diagrams are indeed objects of graphical space even though they are not yet representations of geometric objects. This graphical space is the place for representations of material objects, for positions and for movements in the tangible physical space, and also the place for representations of geometrical objects. In one case, it is a question of models for objects and actions, and in the other case, for "physical" representations of theoretical objects and relations. The distinction between drawings and figures made by a number of researchers (Parzysz 1988; Laborde and Capponi 1994) is significant when seen from this point of view. Moreover, the objects of graphical space can be studied for themselves, and in this case, the domain of reference is not outside of the graphical space itself. This triple domain of reference for graphical objects can be a difficulty for pupils, as shown in the work of Duval (1994, 1995). The objects of the graphical space are representations, between which we can include linguistic expressions.

The role of language and/or semiotic tools in the teaching of geometry has been studied by many research studies in France, such as that of Duval (1995) and Bulf, Mathé and Mithalal (2014). Moving from "seen" to "known" as Parzysz (1988) indicates involves a language work of formulation and validation, as shown by the work of Brousseau (2000), and more recently, by Mathé and Mithalal-Le Doze (2019). These researchers identify four language-learning issues:

> Moving from the instrumented analysis of drawings to the formulation of a geometric analysis of figures, implies being able to: i, name the geometric objects mobilized, figurative units of dimensions 2, 1 or 0 of the complex drawing under study; ii, name relations between these geometric objects independently of the instruments that carry them; iii, articulate mathematical discourse and the organization of the geometric analysis of the drawings; iv, move from a pragmatic description of a figure to an analysis that allows it to be characterized (defined) geometrically (op. cit., p. 69, *translated by author*).

The role of visualization has also been studied in relation to what we see or do not see in a drawing (figure). Duval (2005) talks about iconic and non-iconic visualization, and emphasizes the importance of dimensional deconstruction (e.g. moving from surfaces to lines and points) to identify geometric objects and their relationships. Other works, such as those of Marchand (2009), have added a classification of relations to space to Van Hiele's taxonomy, and highlight the importance of mental images (Marchand 2020). The role of the body and gestures is also studied by research studies that focus on the mediation of the body in learning. Celi's work (in this volume) focuses on this bodily dimension in the wake of work on the use of templates (Coutat and Vendeira-Maréchal 2015). Soury-Lavergne and Maschietto (2019), in their overview of research spanning the last 10 years, indicate that "the specificities of early geometry learning, especially in primary school, have not been truly addressed" (p. 100, *translated by author*). They then show how digital technologies, "the spatial and the geometric with technology", can have a role in preliminary learning, by articulating them to spatial knowledge. We refer the reader to this article. Let us just emphasize that technology, and in particular DGE, "provides an augmented graphical space that leads pupils to work at the level of the figure and not only of the drawing" (ibid., p. 103, *translated by author*), at the level of the possible treatments of these drawings (or figures), and also at the level of the feedback that it allows (ibid.). These authors propose that the digital be associated with other tools and objects of the tangible-tangible space, what they call an "artifact duo" (tangible artifact and digital artifact). Other researchers have focused on the relationship between space and geometry as it relates to Dynamic Geometry. Sinclair and Bruce (2015) provide examples of work that focuses on the relationship between the visual and kinesthetic within geometry learning, such as Battista (2008). She also emphasizes the importance of dragging in drawing out invariants, and in paying attention to those invariants, as discussed above.

These results are far from exhaustive, but they do allow us to justify the principles that are the basis of the design for the experimental situations that we present next.

6.3. Designing situations for first grade primary school

The ERTé MAGI (Mieux Apprendre la Géométrie avec l'Informatique: Better Learning of Geometry with Computers), a program financed by the Direction de la Recherche et la Direction de la Technologie, was led by Colette Laborde between 2003 and 2006, and involved members of the IUFMs[1] of Grenoble, Aix-Marseille, Amiens and Versailles. The aim of this project was to design situations, experiment them in class, analyze them and produce resources (in the form of a CD-ROM) for teachers. In this project, several teams were involved, and we were part of the Toulon-Aix team, composed of three members: a first-grade teacher (first year of primary school), an ICTE trainer and a teacher-researcher. In this context, we designed and experimented with a certain number of situations intended for pupils of the CP class in France (6-year-old pupils). This old work has not been published. Why come back to it? As Soury-Lavergne (2020) says, the potentialities of DGE have not truly been explored, in particular for primary education. Our research on this environment has been placed at this level of education, and I think it is important to recall and disseminate it, given that it is not widely known. Moreover, research on the use of DGE with 6-year-old pupils is not extensive. It would therefore be a shame to not publish, especially since we have been presenting them regularly in the context of mathematical training for primary school teachers. A final reason is that this work is in line with what is indicated in more recent publications, namely, the need for new research on the uses of Dynamic Geometry to work on the relationship between tangible–tangible space and geometric space. Therefore, our old research seems to be still relevant and can be looked at again in light of more recent works. We will present the situations we have experimented with and the principles underlying their conception.

---

1 Institut Universitaire de Formation des Maîtres (teacher training institute).

6.3.1. *Our choices for designing situations*

A first choice is related to the instrumental dimension: we consider that instrumental genesis is a complex process and that it must be organized by the teacher. This principle is consistent with a number of research studies (Soury-Lavergne and Maschietto 2019).

Part of the session is devoted to organizing the pupil's first encounter with some of the software's features. Since movement is an essential element of the Dynamic Geometry Environment, we feel it is important to begin with a motor activity that allows pupils to manipulate the software by moving points and objects (circles). We want them to observe and learn the gestures necessary for these movements. We take advantage of this motor activity to have the pupils describe the static figure (red and blue circles, large and small) and spatial relationships (outside, inside). Given the level of the class, we will not insist much on the construction of geometric objects, but we will still propose to the pupils to construct objects such as the segment, the circle and the triangle. A second choice is related to the institutional dimension: we want the teaching scenario (set of situations proposed to the pupils) to take into account the programs and the official instructions. Therefore, the types of tasks, such as describing a figure or reproducing a figure, will be proposed. A third choice relates to the praxeological dimension. In this regard, it seems necessary to us, to interweave, from a given point, the types of tasks of the DGE and the types of paper-and-pencil tasks. If we want the software to be integrated into the pupils' work, it seems important to us that the DGE becomes a tool with the same status as other tools, and that the work is organized in such a way so as to think about the articulation between the different systems of tools and spaces. Here, we find the idea of the "artifact duo" (Soury-Lavergne and Maschietto 2019). These sessions in the micro-space were articulated by sessions in the meso-space (Gobert 2001).

6.3.2. *Presentation of situations*

We built our scenario on eight situations, some in microspace with the Cabri-geometer software as well as paper-and-pencil, others in meso-space. We present in Table 6.1 these situations. We describe here only four of these situations with Cabri, which will allow us to show the uses of the software to work on the relationship between tangible space and geometric space. The sessions last about 40 minutes each, and the work is done in half-class batches (about 10 pupils in each group). The pupils work in pairs, mainly due to material constraints (number of computers). We describe each situation with a summary, the starting or target figures, and indicate the types of tasks (designated, respectively, by Ti, Tg and Ts, for instrumental, geometrical and spatial tasks) as well as the working methods. The four situations (Situation 1: Blue and red circles; Situation 2: Little ants; Situation 3: Doll; Situation 4: Compositions) are presented in Appendix 1.

	Micro-space		Meso-space
Cabri	**Paper-and-pencil**		
Blue and red circles		Draw me a route	Hoops: paths between two, three, four hoops
Ants		Draw me a route	Movements of various shapes
Doll			
Composition of figures (reproductions)	Reproductions (templates)		
Grid reproductions	Grid reproductions		

Table 6.1. *Summary of situations*

6.4. Analysis of the situations for the first-grade class

We analyze these situations from the point of view of the instrumental and praxeological modes of integration, by highlighting the links between the mathematical dimension, the instrumental dimension and the graphical dimension. We will thus be able to show the uses of DGE to establish bridges between tangible–tangible space and geometric space.

6.4.1. *Instrumental dimension: perceptive–gestural level*

The first level of the instrumental dimension is the *material*, *gestural* and *perceptive* level.

The software is an object of the tangible–tangible space, and also an object of the graphical space. The instrumental initiation takes into account this perceptive–gestural level. Dragging, as research shows, is essential in DGE (Restrepo 2008). The task type Ti1 (moving points and objects) is thus present in several situations (notably Situations 1 and 2). Objects move, but how do they move? The pupils must make gestures with the mouse and be attentive to the feedback from the software by perceiving the effects of these gestures. For example, when the pupils want to move the circles (Situation 1, Appendix 1), they manipulate the mouse to get closer to the circle: the software writes "circle" as feedback.

Next, to move the circle, the pupil must bring the mouse closer to the center (the circle had been built with a fixed radius), and the pointer turns into a small hand which, when closed, allows the pupil to grab the circle and to move it. The work on gesture and perception is essential. One pupil describes this relationship between gestuality and perception as follows: "I took the mouse, I put the hand under the point and I was waiting for the hand, it closes to bring it to the red circle, and I did the same for the blue circle."

Some pupils had difficulty observing this gesture–perception relationship because they did not observe the effects of the mouse gestures and correlate them to what they saw on the screen: there was a disjunction between the gesture with the mouse and the perception of the drawings and their vertical movement. Other pupils also had difficulty establishing these relationships because they were not waiting for the feedback given by the software. Faced with these difficulties, the teacher highlighted the relationship between gesture and perception during the sharing of information phase. To do this, she called upon semiotic tools specific to the software: the transformation of the pointer into a hand was pointed out by the pupils and taken up by the teacher. The presence of the open hand and the hand that closes became indications that the object can move. This perceptual–gestural level is also present when drawing geometric objects with the software. Therefore, to trace a segment, the teacher indicates to the pupil that they must "activate the segment" and then make two clicks: "First click, we move the mouse and a second click". This way of telling the pupil how to draw a segment emphasizes the gesture and what the gesture will allow the pupil to do and see. The construction of the triangle is presented by the icon and three clicks, and that of the circle by two clicks (the center and a point of the circle). The graphical space is thus summoned by the icons (of the software), by the clues (written and visual) and by the drawings.

6.4.2. *Instrumental dimension: spatial–geometric relationships*

During the instrumental initiation, we aim at instrumental knowledge linked to the functioning of the software, which is the case with the three situations (circles, ants and dolls) that were proposed to the pupils. Situation 2 (Appendix 1) poses the problem: where do the points move? What can be observed about these movements?

This situation allows for an encounter with the three types of "cabri points": the free point, which moves everywhere; the constrained point, because it belongs to another object and moves within that object and the point dependent on an object (e.g. the middle of a segment), because it moves if and when the object is moved. The question about the places where the points move allows us to move beyond the perceptual–gestural dimension.

Indeed, these three types of point indicate relations of belonging, the point of the plane, the point belonging to an object (segment, for example), and the point of intersection between two or more objects, or the point defined by an intrinsic relation with an object (middle of a segment). These relations of belonging make it possible to move from the spatial to the geometric, to move from the action "move where?" to a geometric relation. In the same way, in the "ants" situation, the pupils have to move the different points that are visible on the screen. By activating the "trace" function, the trajectories of these points become visible. However, the choice of construction of these points was made in such a way so as to describe particular trajectories: the point that cannot be moved or that can go anywhere (borderline cases), the points that describe a trajectory having the shape of a segment, a square or a circle. These objects obtained as trajectories of points are objects of the graphical space of shapes, but shapes that visually refer to geometric figures. The global recognition of the shape is a step in the recognition of a geometric figure, that of the iconic visualization (Duval 2005).

This situation is associated with two other situations, one in micro-space where pupils had to classify shapes, and the other in meso-space where pupils had to move from point A to point B, in different ways (circular, rectilinear or mixed movements), and then they had to draw these paths on a sheet of paper. Most of the pupils in these first-grade classes recognize the usual shapes and their names, but recognizing them as trajectories is an entirely other matter. The dynamic (the trajectory, the movement) is associated with the static (the overall shape), and thus gives two points of view on the shape that will become a geometric figure. We are here at a second level of the instrumental dimension where the instrumental is always associated with the spatial (the movement, the visual and the gesture) but is also associated with the geometrical (relations of belonging, usual geometrical forms and figures as trajectories of points): spatio-geometrical level (Berthelot and Salin 1992).

6.4.3. *Instrumental dimension: exploration and graphical space*

Situation 3 (Appendix 1) is organized into two stages: the first stage is an instrumental exploration stage; the second is a stage of reproducing the shape. In the first stage, the teacher asks what they see: a doll, the body, the head, the arms and the legs. And then, she asks if these elements of the doll look like something. Pupils are able to talk about "circle", "lines" and "triangles".

The teacher then talks about "circle", "line segment" and "triangle", but does not emphasize the use of this vocabulary. Pupils explore this object in graphical space by moving all the points that can be moved. They observe how these objects move: enlargements, reductions (circle, segments) and deformations (a triangle that flattens or stretches). This allows them to see that shapes do not always remain the same size, but that they can enlarge or reduce while maintaining an invariant geometric shape (the enlarged circle is still a circle). In addition, this exploration also allows pupils to identify geometric shapes, the basic figurative units that make up the doll. The teacher concludes this step by doing a group conclusory session on the different ways of moving an object and on the different shapes that make up the doll. She emphasizes the different gestures and feedback from the software for these movements: the hand, the red dot and the different movements. She then returns to the different shapes that make up the "doll" by telling them that they are going to "draw" their own doll.

The second stage is a mixed stage (instrumental and graphical): it is a question of learning to draw segments, circles and triangles with the software, and to reproduce the graphical shape (doll). The perceptual–gestural level is still very present to trace graphical objects with the software. Therefore, to draw a segment, the teacher indicates to the pupil that he must "activate the segment" and then make two clicks: "first click, one moves the mouse and then a second click". This way of telling the pupil how to draw a segment emphasizes the gesture and what the gesture will make it possible to do and see. The construction of the triangle is presented by the icon and three clicks, and that of the circle by the icon and two clicks (the center and a point of the circle). The graphical space is thus summoned by the icons (of the software), by the clues (written and visual) and by the drawings.

The pupils are then left to reproduce and create their doll themselves, the criteria for success being to have a doll with all its elements. The goal is not to reproduce the original doll with the same dimensions or for it to have symmetrical objects (such as arms and legs), but to create the same basic elements and some of their relationships.

These objects of the graphical space refer to objects of the tangible space (a doll) and also to geometrical objects. From this point of view, this doll corresponds to a double model, of the real and of the geometric space. The doll allows for instrumental initiation through the creation of objects, as the pupils had not yet created "cabri" objects (before the figures were given), and allows for the link with geometry (complex figure composed of simple figures) which will be developed in the "compositions and reproductions" situation.

6.4.4. *Instrumental dimension: tool-geometric space symbiosis*

In Situation 4 (Appendix 1), the tasks aimed at are of a geometric type: reproduce a complex figure. This type of task is usual in primary school, especially with paper and pencil and using the usual tools. The instrumental dimension here is the symbiosis between the tool and the geometric space (at primary school level). The type of mathematical task is what drives this, and what will change are the techniques used depending on whether the software or other tools (such as templates) are used. The type of instrumental task (moving objects) is at the service of the mathematical task which, in principle, is no longer problematic. In reality, this is not the case for all pupils as we have observed. Some pupils still have difficulty moving the graphical objects and putting them in the right positions in relation to each other, especially since in some cases the starting objects had to be enlarged or reduced and adjusted to get the right configuration. This difficulty is not present when the pupils are given templates: in this case, they can choose the template according to the "size" they need. This situation is associated with a paper-and-pencil situation in two ways: the figure is given in paper-and-pencil and must be reproduced with the software; conversely, the pupil creates a figure with the software and must reproduce it using templates. The aim is not to make a reproduction by creating a figure that can be superimposed on top of the original figure, but to create a figure that respects the relative positions of the different basic figures.

This work allowed them to break down a complex figure into figurative units and to identify the relative positions of these units. During the sharing phase, the teacher emphasized the fact that many different complex figures can be made with the same three basic figures, and insisted on the relative positions of these figurative units. This level of the instrumental dimension is a geometric level where the tool is interwoven with the mathematical dimension: the reproduction of a complex figure taking into account the constraints of the tool (graphical space) make it possible to pose the problem of recognizing the basic figures and the relative positions of these figures. Let us now look at the praxeological dimension.

6.4.5. *Praxeological dimension*

Table 6.2 summarizes the different types of tasks encountered only in situations where DG is involved. Table 6.3 shows the organization of these types of tasks in relation to the three spaces (tangible, graphical, geometric) and the instrumental dimension. It provides us with several pieces of information.

The first observation is that task types that might be seen as belonging exclusively to a certain dimension or space can be associated with other spaces. A second observation is the interweaving of the different spaces in relation to the same types of tasks. From these two observations, we can see the praxeological continuity of the three spaces for these types of tasks. Moreover, all the proposed tasks are related to the graphical space, which shows the pivotal role of this space in relation to the other two spaces (Perrin-Glorian and Godin 2018). The DGE as an object of the graphical space thus makes it possible to make a link between the tangible space and the geometric space, for example, when geometric shapes are seen as trajectories of points. DGE is also a tool, and as such, has its own functionalities, hence the need for an instrumental initiation. As a tool, it imposes specific constraints on different types of geometric tasks, such as reproducing a complex figure. Hence, the transition to an instrumental symbiosis. We are in a mode of praxeological organization of interweaving between different types of tasks and techniques, according to the tools and spaces (paper-and-pencil, Dynamic Geometry). It is not only a question of a complementary instrumental genesis between the different artifacts ("artifact duo" in the sense of Soury-Lavergne and Maschietto), it is a question of thinking of the whole of the praxeological organization as a mode of intertwining, even a maximal mode. In our study of first-grade situations, this maximal mode would imply the presence of strong techniques (Assude and Mercier 2007), namely, techniques associated with a technological discourse (in the sense of Chevallard). However, this is not the case. The techniques present are weak techniques, since they are associated with a discourse that describes the technique. This discourse could be seen as a "geometric technical language" (Petitfour 2017) but not only, as we will see later, with the task genre "describe". Genre and type of task are taken, here, in Chevallard's sense: thus "describe" is a genre of task while "describe a geometric figure" is a type of task.

Types of instrument tasks	Ti1: Move points and objects. Ti2: Use the trace to visualize trajectories. Ti3: Recognize the different "cabri" points (free point, constrained point and dependent point). Ti4: Use the menus to trace objects.
Types of spatial tasks	Te1: Identify spatial relationships (inside, outside).
Types of geometric tasks	Tg1: Describe a simple figure. Tg2: Use appropriate geometric vocabulary. Tg3: Recognize geometric figures as paths of points (segments, circles, square). Tg4: Describe a complex figure by recognizing simple geometric figures (segment, circle, triangle). Tg5: Trace segments, circles and triangles. Tg6: Reproduce a complex figure from simple figures.

Table 6.2. *Types of tasks*

Tool	Tangible space	Graphical space	Geometric space
Ti1	Ti1	Ti1	
Ti2	Ti2	Ti2	
Te1	Te1	Te1	
		Tg1	Tg1
		Tg2	Tg2
Ti3	Ti3	Ti3	Ti3
Tg3	Tg3	Tg3	Tg3
Ti4	Ti4	Ti4	Ti4
	Tg4	Tg4	Tg4
Tg5		Tg5	Tg5
Tg6		Tg6	Tg6

Table 6.3. *Types of tasks by types of space*

6.4.6. *Praxeological dimension: observe and describe*

The task type "describe" is broken down into task types for three out of four of the situations presented here. The task type "observe" does not appear explicitly, as it is implied in each situation as it is an essential element of the work economy.

Each session begins with observation, since this is an essential step in what the pupil must do. The first function of observation is to highlight the gesture–perception relationship, as already mentioned above. A second function is to install certain rules of the teaching contract. The pupils are asked by the teacher to describe what they observe, because the work with the software also aims to establish a certain relationship to geometry: to observe in order to describe what is observed. We give here an example of the first session (Situation 1). The teacher began by asking the pupils: "What do you see?" The pupils gave answers such as: "Red circles, a blue circle", "Rounds" and "Small red circles and small blue circles", which correspond to the visible objects, emphasizing the vocabulary. In addition, the pupils indicate "There are dots" (centers of the circles) or "There are crosses" or "This is the mouse" or "It says this circle", which corresponds to one of the feedback comments of the software. In the discussion, the teacher emphasizes the importance of the observation itself, the fact that the pupils must pay attention to the gesture–perception relationship with regard to the software feedback, and also the fact that they must describe what they observe.

The use of the trace in the first two situations (circles and ants) also helps to introduce this observation work in the classroom. The observation of trajectories is a way to show the pupils that the shapes, which they recognize globally, can also be seen as trajectories of a point that moves according to certain constraints. This work is not usually done with paper and pencil, but the software makes it possible to do it in connection with the "point on object" in a contextualized way. Situation 2 was presented as follows: "Little ants are looking for food. They have to go to the point to get food and come back home. We are going to help the little ants move around to get food, but they can't go everywhere. How far can each ant go? You will start with ant A, then ants B, C, D and E in order. Will all the ants have something to eat?" This contextualization remains in the memory for some pupils, as this pupil refers to the following session: "We had little ants and we had to make them eat."

This situation caused some difficulty for the pupils because of the construction choices. One of the difficulties was that the pupils wanted to move the ants by the letter and not by the dot. Adjustments were made during the session with the second group, who therefore did not encounter these difficulties. Observation is thus a means of installing certain rules of the teaching contract, such as:

– observe to establish relationships, to identify invariants;

– observe and interpret feedback from the software;

– observe trajectories and recognize geometric shapes;

– observe to describe and record.

In the first-grade class, the description work was done orally with three objectives:

– Describing to identify objects, for example, "There are big circles and little circles" or "There are little red circles and blue circles."

– Describing actions, for example, "There were little circles and big circles and you had to put the little ones in the big ones" or "There were circles, the green circle made a line, the red circle made a square, the black circle made squiggles, the blue circle made a circle."

– Describe relationships, for example, the relative positions of figurative units in a complex figure: "The square is inside the triangle and the circle is outside."

The type of task "describe" is present in the first-grade curriculum, but Dynamic Geometry adds to it, in the sense that the description must take into account the evolution of the drawings when the points or objects are moved, as shown in the work cited on the importance of dragging. The description takes into account a static state that may be different from another static state.

6.5. Conclusion

We first presented research results on the uses of DGE that are old but not well known. These results seem to be confirmed by more recent works, which can be seen as an indication of their robustness.

Next, we presented a set of situations for cycle 2 of the French primary school system, whose analysis was not published at the time. Given the lack of work on DGE at this level of education, their dissemination seemed relevant to us. The design principles of these situations are based on research results and the analysis framework is made up of four dimensions: the tool system and the three spaces, as we can summarize in Figure 6.3. The arrows indicate all the relationships between these dimensions; the two triangles indicate that the tool system and the graphical space are two fundamental bases in the study of the relationships between tangible space and geometric space.

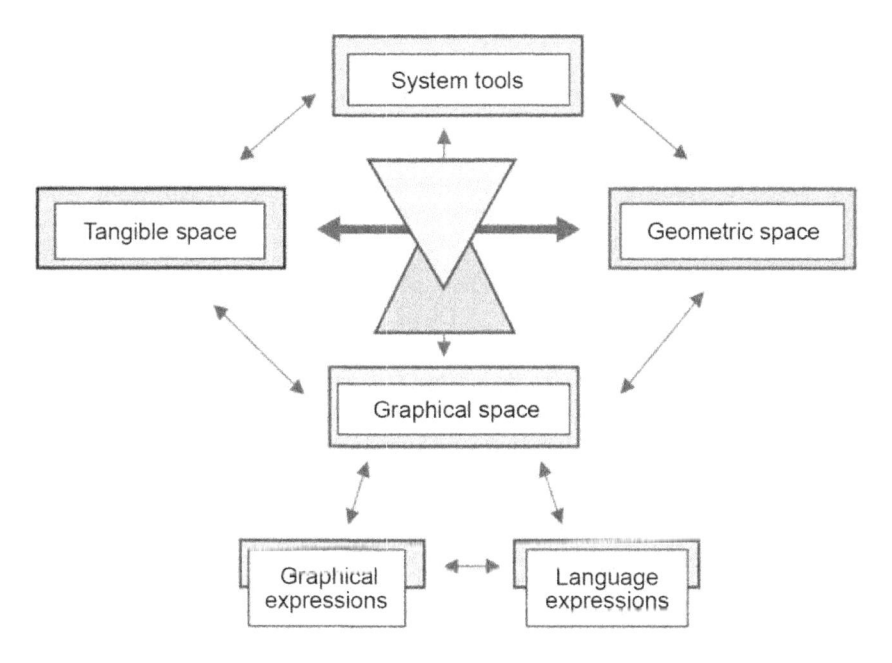

Figure 6.3. *Relationships between the four dimensions*

The modes of instrumental and praxeological integration allowed us to show that the design of the situations was aimed at the mode of interweaving task types and techniques in relation to these four dimensions, as well as an evolution from instrumental initiation to instrumental symbiosis. The pivotal role of graphical space was highlighted by following other research (Mathé and Mithalal 2019; Perrin-Glorian and Godin 2018). Given the importance of graphical space when using DGE, we want to end this chapter by posturing some questions.

The first question is: "What are the objects of graphical space?" In a general way, we could consider several types of objects, in particular linguistic expressions (all the discourses necessary to describe, formulate, validate, justify techniques, etc.) and graphical expressions (drawings, diagrams, pictograms, charts, etc.). This denomination is analogous to that of algebraic expressions, in the algebraic framework, and designates the way of expressing something. A graphical expression (like the algebraic expression) expresses something. The significance of this general designation is that it allows, on the one hand, to put together a certain number of terms, and on the other hand, it can open the way to "operations". In the case of DGE, a drawing expresses an object which can be of tangible space (e.g. a diagram of the movement of a person from point A to point B) or expresses an object of geometric space (a geometric figure) or an object of graphical space itself (a Cabri-drawing, for example). We thus encounter the problem of the double or triple reference (Duval 1995).

Duval's work on the three cognitive operations of semiotic representation, which are the formation, processing and conversion of a representation, can be the starting point for operations on graphical expressions, but by focusing on the materiality of the operations: "How is a graphical expression formed? What processing can be done on graphical expressions? What conversion into another register of representation or another framework?" Duval studied this operation, that is, the treatment of a particular graphical expression (drawing or figure): the cuttings and the arrangements, the dimensional deconstruction are operations on these graphical expressions. The formation of a graphical expression is also an important problem. In the case of DGE, we have seen that the formation of these graphical expressions, which are objects or drawings, obeys software constraints, and then, the way the user uses the software. We have seen examples of instrumental initiation where pupils draw objects under constraints, for example, point on object. The treatment of these graphical expressions is also an essential element, particularly in terms of dragging. We will not revisit here the different functions of dragging in DGE (Restrepo 2008). Other treatments are also possible, for example, associating a measurement to the length or the area of a figure. The purpose of these treatments is to express something about the object. The study of graphical expressions and the "operations" that can be done with them could be developed, following the example of much work in geometry teachings that is interested in language expressions (for a synthesis, see Mathé and Mithalal-Le Doze 2019). Following Duval, these researchers work along these lines as they speak of instrumental deconstruction, mereological decomposition or dimensional deconstruction as "essential operations for learning geometry" (p. 58, *translated by author*).

Generic questions could be asked: What operations could be performed with such a graphical expression? What does the result of these operations express about the object itself (knowing that the object is relative to a domain of reference, for example, the tangible space or the geometric space)? Are two graphical expressions "equivalent"? For example, in the case of DGE, are two drawings "equivalent"? Under what conditions would they be "equivalent"? As we have just seen, these questions are in line with the work of Duval and other researchers.

6.6. References

Abboud-Blanchard, M. (2013). Les technologies dans l'enseignement des mathématiques. Études des pratiques et de la formation des enseignants. Synthèses et nouvelles perspectives. Note de synthèse pour l'Habilitation à Diriger des Recherches, Université Paris Diderot, Paris.

Abboud-Blanchard, M. and Rogalski, J. (2017). Des outils conceptuels pour analyser l'activité de l'enseignant ordinaire utilisant des technologies en classe. *Recherches en didactique des mathématiques*, 37(2–3), 161–216.

Arzarello, F., Olivero, F., Paola, D., Robutti, O. (2002). A cognitive analysis of dragging practices in Cabri environments. *Zentralblatt Für Didaktik Der Mathematik*, 34(3), 66–72.

Assude, T. (2005). Time management in the work economy of a class. A case study: Integration of Cabri in primary school mathematics teaching. *Educational Studies in Mathematics*, 59(1), 183–203.

Assude, T. (2007). Teachers' practices and degree of ICT integration. *Proceedings of the Fifth Congress of the European Society for Research in Mathematics Education*, 1339–1348.

Assude, T. and Gélis, J.M. (2002). Dialectique ancien-nouveau dans l'intégration de Cabri-géomètre à l'école primaire. *Educational Studies in Mathematics*, 50, 259–287.

Assude, T. and Loisy, C. (2009). Plus-value et valeur didactique des technologies numériques dans l'enseignement. Esquisse de théorisation. *Revista Quadrante*, 18(1/2), 7–28.

Assude, T. and Mercier, A. (2007). L'action conjointe professeur élèves dans un système didactique orienté vers les mathématiques. In *Agir ensemble. L'action conjointe du professeur et des élèves*, Sensevy, G. and Mercier, A. (eds). Presses Universitaires de Rennes, Rennes.

Assude, T., Capponi, B., Bertomeu, P., Bonnet, J.-F. (1996). De l'économie et de l'écologie du travail avec le logiciel Cabri-géomètre, *Petit x*, 44, 53–79.

Battista, M.-T. (2008). Development of the shape makers geometry microworld. In *Research on Technology and the Teaching and the Learning of Mathematics: Vol. 2*, Blume, G.W. and Heid, M.K. (eds). Information Age, Charlotte, NC.

Berthelot, R. and Salin, M.-H. (1992). L'enseignement de l'espace et de la géométrie dans la scolarité obligatoire. PhD Thesis, Université Bordeaux 1, Bordeaux.

Bloch, I. and Pressiat, A. (2009). L'enseignement de la géométrie, de l'école au début du collège : situations et connaissances. In *Nouvelles perspectives en didactique des mathématiques*, Bloch, I. and Conne, F. (eds). La Pensée Sauvage, Grenoble.

Brousseau, G. (1983). Étude des questions d'enseignement. Un exemple : la géométrie. Séminaire de didactique des mathématiques et de l'informatique, Université Joseph Fourier, Grenoble.

Brousseau, G. (2000). Les propriétés didactiques de la géométrie élémentaire. *Actes du séminaire de didactique des mathématiques*, Université de Crète, Rethymon, 67–83 [Online]. Available at: https://hal.archives-ouvertes.fr/hal-00515110/fr/.

Bulf, C., Mathé, A.-C., Mithalal, J. (2014). Apprendre en géométrie, entre adaptation et acculturation. Langage et activité géométrique. *Spirale – Revue de recherches en éducation*, 54, 29–48.

Chevallard, Y. (1992). Intégration et viabilité des objets informatiques dans l'enseignement des mathématiques. In *L'ordinateur pour enseigner les mathématiques*, Cornu, B. (ed.). P.U.F., Paris.

Chevallard, Y. (1999). L'analyse des pratiques enseignantes en théorie anthropologique du didactique. *Recherches en didactique des mathématiques*, 19(2), 221–266.

Chevallard, Y. and Jullien, M. (1991). Autour de l'enseignement de la géométrie au collège (première partie). *Petit x*, 27, 41–76.

Chevallard, Y. and Mercier, A. (1987). *Sur la formation historique du temps didactique*. IREM d'Aix-Marseille, Marseille.

Coutat, S. and Vendeira-Maréchal, C. (2015). Des pointes, des pics et des arrondis en 1P-2P. *Math-école*, 223, 14–19.

Duval, R. (1994). Les différents fonctionnements d'une figure dans une démarche géométrique. *Repères IREM*, 17, 121–138.

Duval, R. (1995). *Sémiosis et pensée humaine*. Peter Lang, Bern.

Duval, R. (2005). Les conditions cognitives de l'apprentissage de la géométrie : développement de la visualisation, différenciation des raisonnements et coordination de leurs fonctionnements. *Annales de didactique et de sciences cognitives*, 10, 5–53.

Emprin, F. (2014). Le point de vue d'ingénieries didactiques. *Actes du 40ème colloque COPIRELEM, Enseignement de la géométrie à l'école. Enjeux et perspectives*. IREM de Nantes, Nantes, 20–31.

ERMEL (2006). *Apprentissages géométriques et résolution de problèmes. Cycle 3*. Hatier, Paris.

ERMEL (2020). *Géométrie CP/CE1. 15 Situations pour l'apprentissage de la géométrie et de l'espace*. Hatier, Paris.

Gobert, S. (2001). Questions de didactique liées aux apports entre la géométrie et l'espace tangible dans le cadre de l'enseignement élémentaire. PhD Thesis, Université Paris Diderot, Paris.

Gousseau-Coutat S. (2006). Intégration de la géométrie dynamique dans l'enseignement de la géométrie pour favoriser la liaison école primaire collège : une ingénierie didactique au collège sur la notion de propriété. PhD Thesis, Université Joseph Fourier, Grenoble.

Healy, L. (2000). Identifying and explaining geometrical relationship: Interactions with robust and soft Cabri constructions. In *Proceedings of the 24th Conference of the International Group for the Psychology of Mathematics Education, Vol. 1*, Nakahara, T. and Koyama, M. (eds). Hiroshima University, Hiroshima.

Hölz, R. (1996). How does "dragging" affect the learning of geometry? *International Journal of Computers for Mathematical Learning*, 1, 169–187.

Houdement, C. (2019). Le spatial et le géométrique, le yin et le yang de l'enseignement de la géométrie. In *Nouvelles perspectives en didactique : géométrie, évaluation des apprentissages mathématiques*, Coppé, S., Roditi, E., Celi, V., Chellougui, F., Tempier, F., Allard, C., Corriveau, C., Haspekian, M., Masselot, P., Rousse, S. et al. (eds). La Pensée Sauvage, Grenoble.

Houdement, C. and Kuzniak, A. (2006). Paradigmes géométriques et enseignement de la géométrie. *Annales de didactique et de sciences cognitives*, 11, 175–193.

Kuzniak, A. (2007). Sur la nature du travail géométrique dans le cadre de la scolarité obligatoire. In *Nouvelles perspectives en didactique des mathématiques*, Bloch, I. and Conne, F. (eds). La Pensée Sauvage, Grenoble.

Laborde, C. (2005). Robust and soft constructions: Two sides of the use of dynamic geometry environments. In *Proceedings of the 10th Asian Technology Conference in Mathematics*, Sung-Chi, C., Hee-Chan, L., Wei-Chi, Y. (eds). Korea National University of Education, Cheong-Ju.

Laborde, C. and Capponi, B. (1994). Cabri-géomètre constituant d'un milieu pour l'apprentissage de la notion de figure géométrique. *Recherches en didactique des mathématiques*, 14(1), 165–210.

Lagrange, J.-B. and Caliskan-Dedeoglu, N. (2009). Usages de la technologie dans des conditions ordinaires : le cas de la géométrie dynamique au collège. *Recherches en didactique des mathématiques*, 29(2), 189–226.

Marchand, P. (2009). Le développement du sens spatial au primaire. *Bulletin AQM*, XLIX (3), 64–79.

Marchand, P. (2020). Quelques assises pour valoriser le développement des connaissances spatiales à l'école primaire. *Recherches en didactique des mathématiques*, 40(2), 135–178.

Mathé, A.-C. and Mithalal-Le Doze, J. (2019). L'usage des dessins et le rôle du langage en géométrie : quelques enjeux pour l'enseignement. In *Nouvelles perspectives en didactique : géométrie, évaluation des apprentissages mathématiques*, Coppé, S., Roditi, E., Celi, V., Chellougui, F., Tempier, F., Allard, C., Corriveau, C., Haspekian, M., Masselot, P., Rousse, S. et al. (eds). La Pensée Sauvage, Grenoble.

Mercier, A. and Tonnelle, J. (1992). Autour de l'enseignement de la géométrie (2e partie). *Petit x*, 29, 15–56.

Parzysz, B. (1988). "Knowing" vs "seeing". Problems of the plane representation of space plane geometry figures. *Educational Studies in Mathematics*, 19, 79–92.

Perrin-Glorian, M.-J. and Godin, M. (2014). De la reproduction de figures géométriques avec des instruments vers leur caractérisation par des énoncés. *Math-école*, 222, 26–36.

Perrin-Glorian, M.-J. and Godin, M. (2018). Géométrie plane : pour une approche cohérente du début de l'école à la fin du collège. *Actes du colloque CORFEM* [Online]. Available at: https://hal.archives-ouvertes.fr/hal-01660837v2/document.

Perrin-Glorian, M.-J., Mathé, A.-C., Leclercq, R. (2013). Comment peut-on penser la continuité de l'enseignement de la géométrie de 6 à 15 ans ? *Repères IREM*, 90, 5–41.

Petitfour, E. (2017). Outils théoriques d'analyse de l'action instrumentée, au service de l'étude de difficultés d'élèves dyspraxiques en géométrie. *Recherches en didactique des mathématiques*, 37(2–3), 247–288.

Rabardel, P. (1995). *Les hommes et les technologies. Approche cognitive des instruments contemporains.* Armand Colin, Paris.

Restrepo, A. (2008). Genèse instrumentale du déplacement en géométrie dynamique chez des élèves de 6ème. PhD Thesis, Université Joseph Fourier, Grenoble.

Ruthven, K. (2009). Towards a naturalistic conceptualisation of technology integration in classroom practice: The example of school mathematics. *Éducation et Didactique*, 3(1), 131–149.

Sinclair, N. and Bruce, C. (2015). New opportunities in geometry education at the primary school, *ZDM*, 47, 319–329.

Soury-Lavergne, S. (2020). La géométrie dynamique pour l'apprentissage et l'enseignement des mathématiques. Report, CNESCO, Paris.

Soury-Lavergne, S. and Maschietto, M. (2019). Connaissances géométriques et connaissances spatiales dans les situations didactiques avec la technologie. In *Nouvelles perspectives en didactique : géométrie, évaluation des apprentissages mathématiques*, Coppé, S., Roditi, E., Celi, V., Chellougui, F., Tempier, F., Allard, C., Corriveau, C., Haspekian, M., Masselot, P., Rousse, S. et al. (eds). La Pensée Sauvage, Grenoble.

Voltolini, A. (2017). Duos d'artefacts matériel et numérique pour l'apprentissage de la géométrie. PhD Thesis, ENS, Lyon.

7

Robotics and Spatial Knowledge

7.1. Introduction

The work described in this chapter deals with the use of programmable floor robots (Figure 7.1) used with French primary school pupils, aged 6–8 years old, in math class.

Figure 7.1. *BeeBot® programmable floor robot*

Since 2015, the official programs concerning the theme "Space and geometry" recommend the use of programmable robots, or programming software, as early as Grades 1, 2 and 3, to work on the skill "To locate (oneself) and to move (oneself) around using landmarks and representations" (Ministère de l'Éducation Nationale 2015). In addition, many works (Komis and Misirli 2011; Grugier and Villemonteix 2017; Romero and Sanabria 2017) report promising results regarding the use of this type of robot by young pupils, whether it is at the level of collaborative problem solving, mastery of a technical object or the acquisition of basic computer programming skills.

Chapter written by Emilie MARI.

For our part, we are interested in spatial knowledge, and the question we are exploring is as follows: "How does the integration of a robot in a mathematics class work on the spatial knowledge of pupils in preparatory school (6–7 years old)?" First, we present a categorization of spatial tasks grouped by genre (Chevallard 1999), based on the definition of spatial knowledge given by Berthelot and Salin (1999) and on the classes of spatial problems elaborated by Fénichel et al. (2004). Next, we present our methodology as well as the test that we have designed, based on these spatial tasks. Finally, we analyze the results obtained for an example of an item that underwent the test.

7.2. Theoretical framework and development for a categorization of spatial tasks

7.2.1. *Spatial knowledge*

We rely on the definition of spatial knowledge given by Berthelot and Salin (1999):

> By spatial knowledge we designate the knowledge that allows a subject an adequate control of their relations to the tangible space. This control is translated by the possibility for them to: recognize, describe, make or transform objects; move, find, communicate the position of objects; recognize, describe, construct or transform a space of life or movement. The mastery of space is the object of learning well before entering school and it continues to develop in parallel ways throughout childhood and adolescence, and even adulthood (op. cit., p. 38, *translated by author*).

This definition, which comes from the field of mathematics education, is in line with the notions discussed in other works related to space and spatial thinking (which we will translate as spatial reasoning), and of which we retain, for our work, the decomposition and recomposition of shapes (Mulligan et al. 2020). Science, technology, engineering and mathematics (STEM) education benefits from the development of spatial thinking, and mathematics in particular is closely linked to this (Wai 2009; Lowrie et al. 2016). Furthermore, "spatial skills are highly malleable and […] training in spatial thinking is effective, durable, and transferable" (Uttal et al. 2013, p. 365), that is, working on spatial reasoning can contribute to pupils' success in mathematics and STEM, as Clements and Sarama (2011) also note.

But in psychology and education, most experiments related to spatial reasoning have been based on psychometric tests (Lowrie 2020). We then sought to develop a test for spatial knowledge, based on a didactic approach. Placing ourselves in the field of mathematics education, we study spatial as a facet of geometry defined by the spatial knowledge that is mobilized in situations that a subject must manage in tangible space (Houdement 2017).

7.2.2. *Types of spatial tasks*

Within the framework of an anthropological approach (Chevallard 1999), we sought to develop a categorization of tasks relating to spatial knowledge, with a view, on the one hand, to being able to situate any exercise proposed to pupils (in textbooks, for example) in this categorization, and on the other hand, to developing a test of spatial knowledge for primary school pupils. This test serves as an indicator to measure the progress of the pupils. We used the classes of spatial problems described by Fénichel, Pauvert and Pfaff (2004), namely: describing a position or displacement, and positioning or moving an object. We have reworked these classes of problems to consider, on the one hand, positioning problems, and on the other hand, displacement problems. We have added problems of understanding geometrical shapes, through the idea of decompositions and recompositions of shapes mentioned previously. These three categories allow us to account for the work required of pupils. As the tasks prescribed to them can be expressed using verbs, we have chosen to group under the label "SEE" the tasks related to perception (locate, identify, etc.); "SAY" the tasks related to a language (say, explain, code, specify, etc.) and "DO" the tasks which imply an action on the part of the pupil (place, move, construct, draw, etc.). We obtain a categorization (see Table 7.1) that we present by task genre, a task genre existing "only in the form of different types of tasks, whose content is more closely specified" (Chevallard 1999, p. 224, *translated by author*).

7.2.3. *Types of tasks and techniques*

We are interested in the techniques implemented to respond to a particular type of task. For each of the types of tasks (Table 7.1) presented and for which we have proposed several possible types of tasks, we have looked for corresponding techniques according to the type of space chosen: the meso space of the classroom or the surrounding space, or the micro-space of the sheet or the screen (Brousseau 2000).

Let us take the task type situate an object and the specific task type situate an object in relation to a landmark outside oneself, which we will denote as T_S^P (with the letter S denoting "situate" and the letter P the problem category "position"). For example, we can list four types of tasks frequently encountered in primary school, which can be broken down according to space: identifying an object in the classroom space and finding an object in the classroom space for *meso-space*; locating an object according to its position on the page or locating a square in a grid for *micro-space*. For each of these types of tasks, several techniques are possible. In the case of locating a cell in a grid, a primary school pupil could, for example, locate the empty or occupied cells for each line, starting from the left to the right of the grid and progressing cell by cell. They could do the same from top to bottom or from bottom to top, or even rely on the proximity of a particular square. This example shows how we have listed, from a categorization by type of task and as exhaustively as possible, the types of tasks that primary school pupils may encounter when working on space. We have thus designed a test with the idea of answering the question: How does the integration of a robot in a mathematics class allow us to work on the spatial knowledge of pupils in their first year of primary school (6–7 years old)?

	Position	Move	Understand shapes
SEE	Locate an object	Recognize a trajectory	Anticipate the result of a move or transformation
SAY	Describe a position	Describe a movement	Name a shape
DO	Place an object in a space	Move an object	Describe a shape

Table 7.1. *Type of spatial tasks*

In the following section, we present the methodology of our research and then, based on the type of task reproducing an assembly, analyze the results of the corresponding spatial knowledge test item.

7.3. Research methodology

We sampled six first grade classes, two of which were in REP[1]. We offered the same test to all classes, in October 2019 and again in January 2020.

Meanwhile, three classes (including one in REP) worked with programmable floor robots (Figure 7.1). The teachers were free to choose how to integrate the robots into their progress. The other three classes worked according to the progression of their textbook, with the aim of getting as close as possible to a real classroom situation. The course of our experimentation is summarized in Figure 7.2.

---

1 In France, REP is a "priority support network" of schools that allows schools located in a disadvantaged social environment to benefit from specific facilities such as smaller class groups.

We now take an example of a test item corresponding to the type of task reproducing an assembly of shapes, and present the results of the test for this item as well as some ideas for analyzing these results.

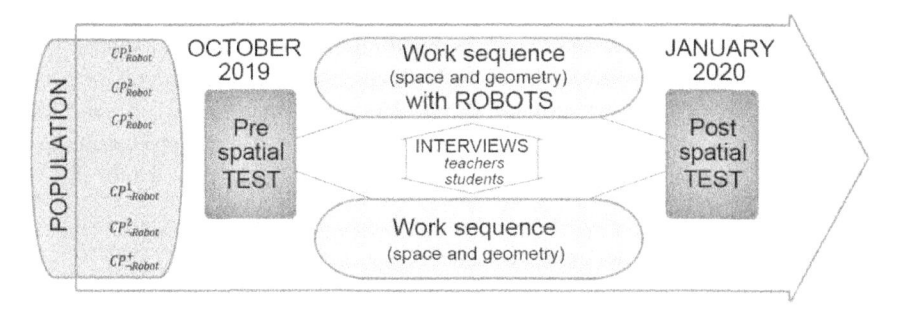

Figure 7.2. *Conduct of the experiment*

7.4. Analysis: reproducing an assembly

7.4.1. *Test item*

For the task type reproduce an assembly (Table 7.1), we propose an exercise in which the pupils would have to reproduce from memory an assembly of simple shapes[2], corresponding to the task type reproduce an assembly of shapes (draw). The shapes are visible for a few seconds (Figure 7.3), then the pupils have to draw them from memory. The aim here is to memorize the shapes of the different elements making up the figure, their relative positions, size and orientation. In order to do this, the pupil can use different techniques, such as associating the shape with a known shape (a house for the first assembly), or breaking down the shape to be reproduced into sub-figures (they must keep the same number of elements that constitute the shape and identify each element). If the pupil breaks down the figure into sub-figures, they must memorize the relative positions of these elements and orient them.

In conjunction with the intervention tool "1, 2, 3... imagine!", from which these forms are extracted, Marchand (2020) distinguishes three levels of abstraction that value spatial knowledge: archaeological, photographic and scenographic. Here, we are situated between the archaeological level (we will see a house, and we will draw the referent that we have of a house independent of spatial knowledge), the

---

[2] The shapes chosen are taken from the intervention tool *1, 2, 3... imagine!* for primary school pupils aged 6–9 years old. © 2018. University of Sherbrooke.

photographic level (we provoke in the pupils a passage towards the mental images of the tangible or geometric objects in play) and perhaps, but it is less likely in this exercise, the scenographic level (we imagine transformations that allow memorization, such as a rotation that would make the large triangle "slide" onto the small square). Pupils can also use techniques related to different kinds of tasks, such as describing a figure. Since the pupil is asked to draw an assembly of shapes from memory, they will draw the shapes one after another in a certain order. They may begin by drawing the triangle, as Pupil 2 does on the pre-test (Figure 7.4), and then erase it before drawing the "house" in one continuous line. These steps can be seen as a sequence of actions, and the timing of the actions could play a role in how spatial knowledge is mobilized.

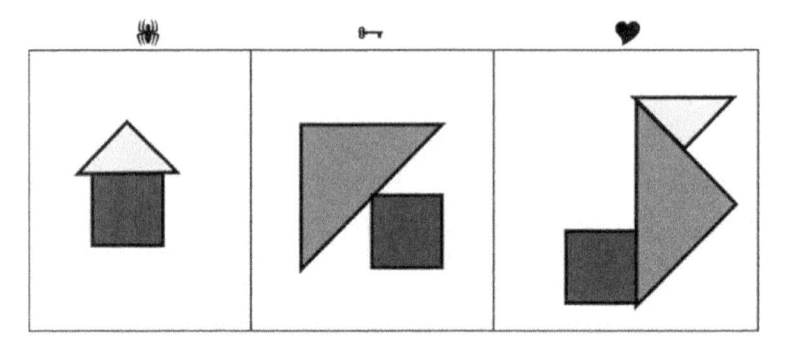

Figure 7.3. *Reproduce an assembly: test item*

7.4.2. Test results

Here are two examples of pupils' achievements for the item reproduce an assembly of figures (Figure 7.4), where the proposed item is identical to both tests. We notice an evolution in terms of the number of elements that compose the figure, their relative position, the orientation or the ratio in size between the elements. These four criteria were used to evaluate the success of the item; a pupil is considered to have succeeded in this item when they have taken into account at least two of these criteria.

The evolution between the pre-test and the post-test is presented in Figure 7.5 according to the classes. A distinction is made between classes that have or have not worked with robots, and the context of the school (REP or not). The results are expressed as percentages of success.

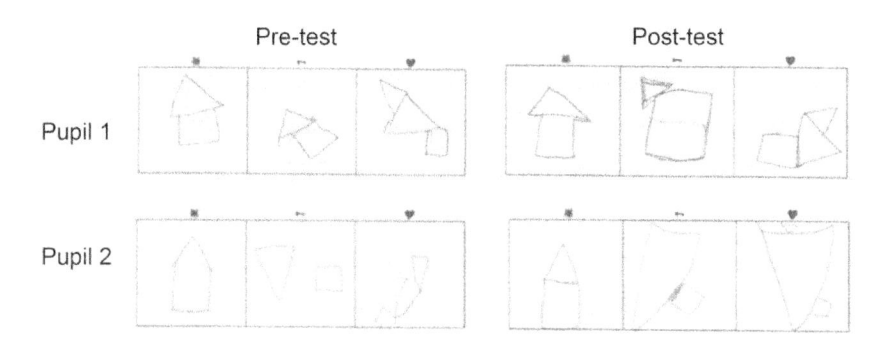

Figure 7.4. *Example of pupils' achievements*

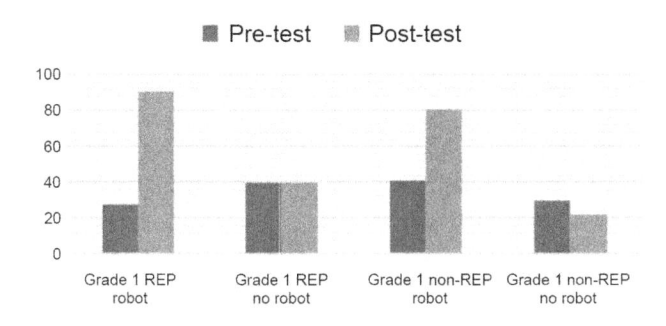

Figure 7.5. *Results for the item: reproduce an assembly*

A significant difference was observed between the classes that worked with robots and the others: the item was more than twice as successful. However, this result needs to be compared with the analysis of what was actually worked on with the robots and in the other classes.

7.4.3. Analysis of the results

In order to better explain these results, we studied the types of tasks addressed by the classes with robots that we were able to film, and by the classes with textbooks. We find three points of interest that we detail in the following sections.

7.4.3.1. *Decomposition of figures into sub-figures*

Once the functions related to the manipulation of the robot are no longer an obstacle, the coding of the robot's movements can encourage the development of sequential thinking, that is, thinking in steps, in chronological order. At first, the pupils followed step-by-step programming and thus had immediate feedback, then the teachers progressively asked them to code in writing all the instructions to be given to the robot so that it followed a given trajectory. This sequential thinking, favoring anticipation before action, could, in our opinion, facilitate the decomposition of figures into sub-figures, an obstacle to learning geometry linked to the mathematization of space (Bishop 1987).

7.4.3.2. Relative positions

The notions of orientation involved when the robot moves, or when working with planes, emphasize the notion of relative position. The pupil who places themselves in the same position as the robot to program its movement will not have to make a left/right inversion (if we want it to turn to the right, we press the right arrow), whereas the pupil who programs a robot while facing it must call upon knowledge about laterality (if we are facing the robot, its left corresponds to our right, and the commands are thus reversed). The pupil who uses a map will be able to orient it to match the surrounding space. A back and forth between meso-space and micro-space techniques could, in our opinion, facilitate the identification of relative positions or even spatial relationships (Marchand, 2020) between different figures within a complex figure.

7.4.3.3. Anticipation and mental images

Finally, the anticipation linked to the robot's movements (the robot's movement is programmed before it is carried out) can make it possible to have greater recourse to the mental images (ibid.) at play while working in geometry. Here again, getting into the habit of imagining the position of the robot at a given moment of its journey is tantamount to anticipating the result of a geometric transformation: a translation in the case of a forward movement, or a rotation in the case of a change of direction.

7.5. Conclusion

We presented a categorization for the types of spatial tasks approached at the level primary school level, a categorization which was then used to carry out a test of spatial knowledge in the field of mathematics education.

We then presented our research on the impact of working with educational robotics on pupils' spatial knowledge using an example of an item, where the results were markedly better for those pupils who worked with robots. However, this is not the case for all the items of the test, and we seek to explain these results by studying more closely the favorable conditions for the use of programmable robots, in order to see whether the hypotheses which emerge here (development of a sequential thought, work on anticipation or relative positions) are confirmed. In order to do this, we are currently developing an observation grid that takes into account the spatial dimension developed in this chapter, the instrumental dimension related to the robot and also the algorithmic dimension that focuses on the programming of the robot.

7.6. References

Berthelot, R. and Salin, M.H. (1999). L'enseignement de l'espace à l'école primaire. *Grand N*, 65, 37–59.

Bishop, A.J. (1987). Quelques obstacles à l'apprentissage de la géométrie. *Études sur l'enseignement des mathématiques*, 5, 1.

Brousseau, G. (2000). Les propriétés didactiques de la géométrie élémentaire. L'étude de l'espace et de la géométrie. *Les propriétés didactiques de la géométrie élémentaire; l'étude de l'espace et de la géométrie*, 67–83.

Chevallard, Y. (1999). L'analyse des pratiques enseignantes en théorie anthropologique du didactique. *Recherches en didactique des mathématiques*, 19, 221–265.

Clements, D.H. and Sarama, J. (2011). Early childhood teacher education: The case of geometry. *Journal of Mathematics Teacher Education*, 14(2), 133–148.

Fénichel, M., Pauvert, M., Pfaff, N. (2004). *Donner du sens aux mathématiques, Vol. 1, Espace et géométrie*. Bordas, Paris.

Grugier, O. and Villemonteix, F. (2017). Apprentissage de la programmation à l'école par l'intermédiaire de robots éducatifs. Des environnements technologiques à intégrer. In *Communication dans l'atelier "Apprentissage de la pensée informatique" lors du colloque EIAH*.

Houdement, C. (2017). Le spatial et la géométrique : le yin et le yang de l'enseignement de la géométrie. In *Nouvelles perspectives en didactique : géométrie, évaluation des apprentissages mathématiques*, Coppé, S. and Roditi, E. (eds). La Pensée Sauvage, Grenoble.

Komis, V. and Misirli, A. (2011). Robotique pédagogique et concepts préliminaires de la programmation à l'école maternelle : une étude de cas basée sur le jouet programmable Bee-Bot. *Sciences et technologies de l'information et de la communication en milieu éducatif : analyse de pratiques et enjeux didactiques*, 271–281.

Lowrie, T., Logan, T., Ramful, A. (2016). Spatial reasoning influences pupils' performance on mathematics tasks. *39th annual conference of the Mathematics Education Research Group of Australasia*, July 1st.

Marchand, P. (2020). Quelques assises pour valoriser le développement des connaissances spatiales à l'école primaire. *Recherches en didactique des mathématiques*, 40(2), 135–178.

Ministère de l'Éducation Nationale (2015). Bulletin officiel no. 17 du 23 avril 2015.

Mulligan, J., Woolcott, G., Mitchelmore, M., Busatto, S., Lai, J., Davis, B. (2020). Evaluating the impact of a spatial reasoning mathematics program (SRMP) intervention in the primary school. *Mathematics Education Research Journal*, 32, 285–305.

Romero, M. and Sanabria, J. (2017). Des projets de robotique pédagogique pour le développement des compétences du XXIe siècle. In *Usages créatifs du numérique pour l'apprentissage au XXIe siècle*, Romero, M., Lille, B., Patino, A. (eds). Presses de l'Université du Québec, Quebec.

Uttal, D.H., Meadow, N.G., Tipton, E., Hand, L.L., Alden, A.R., Warren, C., Newcombe, N.S. (2013). The malleability of spatial skills: A meta-analysis of training studies. *Psychological Bulletin*, 139(2), 352.

Wai, J., Lubinski, D., Benbow, C.P. (2009). Spatial ability for STEM domains: Aligning over 50 years of cumulative psychological knowledge solidifies its importance. *Journal of Educational Psychology*, 101(4), 817.

8

Contribution of a Human Interaction Simulator to Teach Geometry to Dyspraxic Pupils

8.1. Introduction

In primary school and at the beginning of junior high school, knowledge of geometric objects, relationships and properties is built up in situations that require actions with drawing and measuring instruments, using types of tasks such as checking geometric relationships or reproducing or constructing figures. In France, the middle school programs (pupils aged 9–11) specify that:

> Situations involving different types of tasks (recognizing, naming, comparing, verifying, describing, reproducing, representing, constructing) involving geometric objects are favored in order to bring out geometric concepts (characterizations and properties of objects, relationships between objects) and to enrich them. A play on the situation constraints, on the material and instruments made available to the pupils, develops the procedures for handling problems and enriching knowledge (MEN 2018, p. 106, *translated by author*).

Tracing with instruments is therefore an injunction of the school curriculum, and is supposed to contribute to the conceptualization of geometric concepts. However, this teaching method is not suitable for all pupils, and in particular for dyspraxic pupils, who find it difficult to manipulate material. Our research aims to propose alternative ways of teaching geometry, accessible to all pupils, and in particular for those who have difficulties in manipulation, more than difficulties in comprehension

---

Chapter written by Fabien EMPRIN and Edith PETITFOUR.

(which is the case for dyspraxic pupils a priori), in order to meet the objectives of the geometry curriculum.

We first present the general research framework with elements on geometry teaching, on dyspraxia and its consequences in geometry (section 8.2). We then study alternatives for teaching geometry taking into account the handicap of dyspraxic pupils (section 8.3). We next present the possibilities offered by a human interaction simulator for geometric learning (section 8.4). Finally, we present and discuss our first experimental results (section 8.5).

8.2. General research framework

8.2.1. *Teaching geometry*

As we have seen previously, drawing with instruments is a necessary part of teaching. It is supposed to help pupils understand geometric concepts. Research in didactics highlights the importance of work spaces in which the manipulation of concrete objects is at the heart of mathematical work: Geometry I or "natural geometry" is one of the three geometric paradigms for Houdement and Kuzniak (2006), and for Berthelot and Salin (1992); the spatial is linked to the geometric. Furthermore, the role of manipulatives in helping pupils learn mathematics is commonly accepted in most educational settings (Sarama and Clements 2016).

However, this approach to instruction, which uses tools and manipulation, is not appropriate for pupils with dyspraxia, as they lack the skills necessary to complete these academic tasks (Elbasan and Kayihan 2012). Tracing accurately with instruments is an inherent component of curricular expectations which these pupils consistently fail to accomplish, despite having the basic abilities (reasoning, language and memory) that would allow them to work in more conceptual workspaces.

Vaivre-Douret et al. (2011) show that, in a population of 43 children (aged 5–15 years old) with dyspraxia or a developmental coordination disorder, there were notable differences between the percentages of difficulties, in the area of language versus mathematics, and in the ability to perform representational gestures versus the ability to imitate gestures: 28% had learning difficulties related to language, 88% had difficulties in mathematics, especially in geometry or else in correctly establishing arithmetic sums. Only 10% of this population failed to perform activities requiring representational gestures. Fifty-six percent failed in gesture imitation, which plays an important role in teaching how to manipulate geometric instruments.

8.2.2. *Dyspraxia and consequences for geometry*

We use the term "dyspraxic" to refer to pupils with a developmental coordination disorder, manifested primarily in gestural skills. The INSERM[1] expert report (2019) specifies the following criteria from the DMS-5[2] as a diagnosis of this disorder for a subject:

> A. Acquisition and performance of good motor coordination skills is significantly below the level expected for the subject's chronological age, given the opportunities to learn and use these skills. Difficulty is evidenced by clumsiness, as well as slowness and imprecision in the performance of motor tasks.
>
> B. Impairments of motor skills in Criterion A significantly and persistently interfere with chronologically appropriate activities of daily living and impact academic/school performance […].
>
> C. The onset of symptoms dates from the early developmental period.
>
> D. Impairments in motor skills are not best explained by intellectual disability or visual impairment, and are not attributable to a motor neurological condition (INSERM 2019, pp. 2–3, *translated by author*).

Dyspraxic pupils have difficulty manipulating material objects. They are unable to coordinate and perform the fine motor skills required to make instrumental constructions (Petitfour 2018). Their graphical productions rarely correspond to what they are trying to trace. We illustrate their difficulties by relating observations of a dyspraxic fifth-grade pupil who had to reproduce, using a grid, the model figure shown in Figure 8.1.

The pupil makes her tracings with the compass by holding either arm of the compass in each hand (see image Compass Hold No.1), her hands hide her visibility of the model and on the reproduction in progress. The distance is imprecise, the support moves while she starts her tracing, and she stops to erase. She makes a new trace, but this time the compass tip slips. She starts again by holding the compass differently so as to keep in place both the needle and the support at the same time (Compass Hold No. 2), continuing in this way to prevent any visual control of the trace in progress. She repeatedly erases her attempts, sometimes drawing arcs

---

1 Institut National de la Santé et de la Recherche Médicale.
2 *Diagnostic and Statistical Manual of Mental Disorders* (5th edition).

without worrying about the starting point, and overshooting the end point, carried away by her momentum.

| Model | Reproduction | Compass Hold No. 1 | Compass Hold No. 2 |

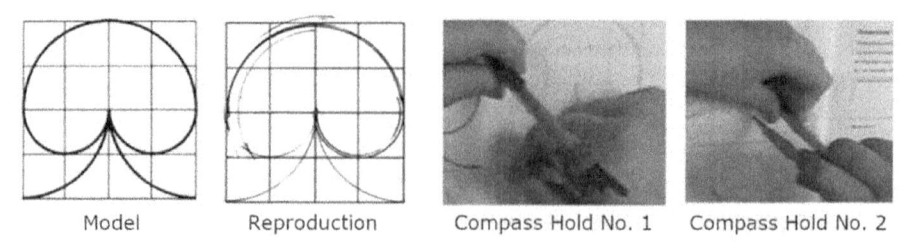

Figure 8.1. *Compass reproduction by a 5th grade dyspraxic pupil*

The pupil's final production shows her trials and inaccuracies in the drawing, even though she was able to identify the centers and radii of the various arcs to be constructed. She spent much more time than the other pupils on this work, in particular because she did not automate the manipulation of the compass like other pupils her age. Her disorder does not allow for this automation, even with intensive practice.

Therefore, the manipulation of instruments constitutes an obstacle for the geometric learning of dyspraxic pupils, because they must focus their efforts on manipulative and organizational aspects to the detriment of the conceptual aspects being targeted.

8.3. What alternatives are there for teaching geometry?

As an alternative to the manipulation of instruments on the paper-and-pencil environment, we first explore the use of digital tools for the realization of geometric drawings, and then we present a device for working in dyads.

8.3.1. *Using tools in a digital environment*

In a digital environment, pupils can make drawings using instructions, pointer/mouse movements or by touching the screen, all of which can be configured and adapted to the disability.

The use of virtual measuring and tracing instruments, such as those in InstrumenPoche[3] developed by Sésamath for work in a digital environment, or such

---

3 https://instrumenpoche.sesamath.net/v2/iep.php.

as those in the GeoTrace[4] Kit (Sagot 2005), developed as part of the national mathematics assessment for pupils with motor or praxis deficits, is visually similar to the use of instruments in the paper-and-pencil environment; the virtual instruments are drawings of tangible instruments (Figure 8.2). This analogous aspect of the instruments allows for some proximity in performing an activity on the computer that their class would do in the paper-and-pencil environment.

InstrumenPoche, Sésamath   Trousse GéoTracé, INSHEA

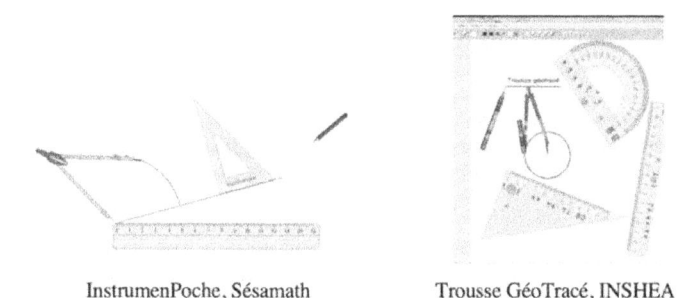

Figure 8.2. *Virtual instruments*

The pointer allows pupils to position the instruments: they can slide them, rotate them, spread the arms of the compass and rotate the compass. The instruments stay in place without the need to hold them.

This has two advantages for users: they can see the construction they are making without their hands interfering with their visibility; they no longer perform a double task when drawing (drawing while holding the instrument), since holding the instrument is no longer their responsibility. However, we have observed some uses of these virtual instruments by pupils, which make transparent the geometric properties at stake because they are supported by the instrument. Here are two examples for a task that consists of constructing a square ABCD from a starting point (Figure 8.3a), with the non-graduated ruler, the set square and the compass, as instruments at the pupils' disposal.

EXAMPLE 8.1.– After measuring the distance AB with the virtual compass, a pupil drags it to bring the compass needle to point C (Figure 8.3b), and then marks point D with the compass pencil tip. The direction of the line (AB) is maintained by the computer as the pupil slides the compass, making the drawing consistent with the expected drawing.

---

4 https://www.inshea.fr/fr/content/trousse-géo-tracé-5-outils-adaptés-de-tracés-géométriques.

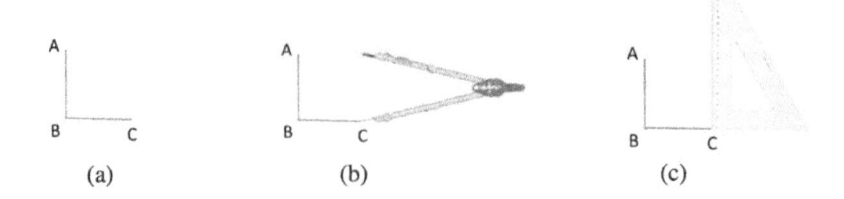

Figure 8.3. *Virtual instruments*

EXAMPLE 8.2.– After selecting the set square, a pupil drags it to bring its right-angle vertex to point C (Figure 8.3c), and then draws the line (DC). The initial orientation of the set square and its maintenance is handled by the computer, not by the pupil. A problematic aspect of these techniques is that they allow the pupil to obtain the desired figure without necessarily being aware of the properties involved, as might be the case in the paper-and-pencil environment. These examples show that this type of instrumental genesis can lead to a lessening of pupils' geometric knowledge.

We have also encountered several limitations in the use of Dynamic Geometry Software (DGS) to be used by dyspraxic pupils as a means of compensation during geometry sessions. The first is related to the screen overload that can result from the construction lines of a figure: pupils can no longer identify the intersection points they are looking for and easily get lost in the various circles and lines traced.

The second is related to the fact that the DGS is a specific constraint system in which the means of feedback and validation are very different from the paper-and-pencil environment. This is illustrated by the following three examples. First, with DGS, an intersection point only exists if it has been created: for example, in addition to having drawn two intersecting lines, we have to create our intersection point, whereas this action does not have to be done in the paper-and-pencil environment. Second, it is not possible to construct the second side of a right angle directly: you must first construct its support line, whereas this is not the case with the square. Finally, a geometric drawing is valid if it resists dragging (Straesser 2002): moving one of the basic elements of the drawing with the mouse deforms it while respecting the geometric properties that were used to draw it, but causes it to lose the apparent spatial properties. This type of validation does not exist in the paper-and-pencil environment. In a classroom situation, a DGS would not help the dyspraxic pupil to perform the same task as the others in the paper-and-pencil environment, but would introduce a different task, requiring specific support from the teacher. However, we have been able to uncover, in teaching practices, various obstacles to the integration of the computer tool in the classroom as a means of teaching geometry to dyspraxic pupils, in the context of school inclusion (Petitfour 2015). We next proposed an alternative solution: changing the nature of the mediation through a peer (Petitfour 2017a).

8.3.2. *Dyadic work arrangement*

We have developed a dyadic work arrangement (Petitfour 2017a) to improve learning opportunities in geometry for dyspraxic pupils, and also for any pupil. This device is based on the language dimension of geometric activity: on the one hand, it exploits the preserved language skills of dyspraxic pupils; on the other hand, it aims to reduce the cognitive leap existing in teaching, between a geometry organized around action with instruments (construction activities or reproduction of figures) and a more discursive geometry (formulation of construction programs, demonstration).

In a task of construction or reproduction of figures with instruments, the work proceeds as follows. The pupils work in dyads, each with a different role: one is an instructor and the other is a constructor. Both follow specific rules (which we will specify later), with a view to implementing an instrumental construction technique leading to the realization of a correct figure (Petitfour 2017a).

We thus qualify a figure constructed with the use of geometrical properties carried by appropriate instruments and resulting in a figure "conforming to the theory". As for the appropriateness of the instruments, for example, the positioning of a ruler in a perceptive way to trace the second side of a right angle from one side does not give a correct figure, even though a good visual perception of the spatial relations leads to a precise tracing (tracing that would be validated if we compared it by superposing it with the expected figure, carried out on a tracing paper). The correct positioning of a set square in order to draw the second side of a right angle will result in a correct drawing. The property "in accordance with the theory" concerns, for example, the respect for the relations of incidence: when we trace the three bisectors of a triangle, we must see a point of intersection and not the small triangle which we can easily obtain with the precision allowed by the instruments.

Let us now turn to the rules to be followed by each of the two members of the dyad and specify some of the basics. The instructor gives verbal instructions to the constructor to make the plots. The instructor is not allowed to manipulate the instruments. They must formulate their instructions using technical language, a language related to the manipulation of the instruments in connection with the mathematical properties at stake in this manipulation. In concrete terms, for a given instrumented action, the instructor first mentions the instrument to be used (1), then gives its position by describing the relations between parts of the instrument and graphical objects already present (2) and finally gives indications about the place of the line, completed, if necessary, by deictic gestures showing where the line should be drawn in relation to the instrument (3). Figure 8.4 illustrates the instructions the instructor might give to draw a right angle at A from a segment [AB]. The result of the corresponding actions of the constructor is in the shaded area of Figure 8.4.

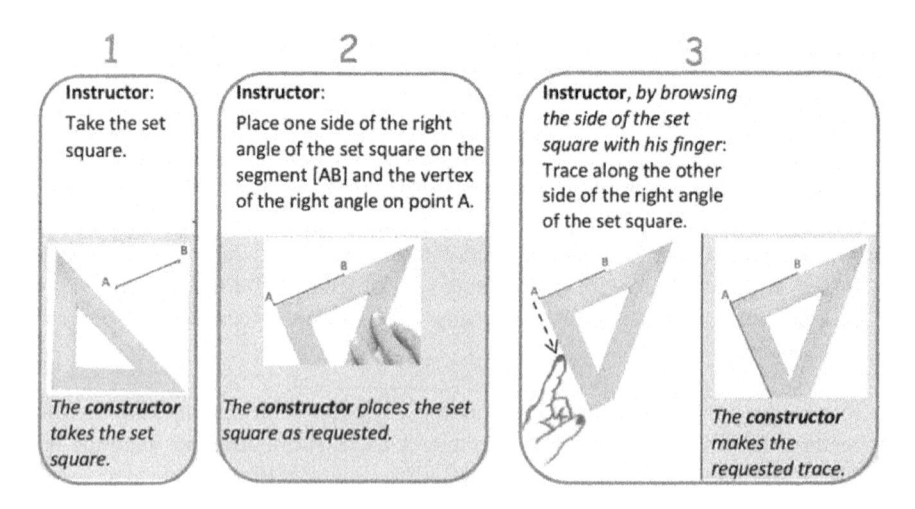

Figure 8.4. *Working in dyad to trace a right angle from a side*

Deictic gestures can be made in step (2) to indicate the plot area, if important, before giving the instructions. In the example in Figure 8.4, the half-plane boundary (AB) in which the plot is to be made will be shown with the hand, if necessary. Deictic gestures (see examples in Figure 8.5) can accompany technical language, but in no way substitute for it. Therefore, the use of deictic terms associated with gestures ("here", "there", etc.) is not allowed. Manipulative guidance such as "move forward a little", "stop", etc., and the use of spatial terms ("up", "left", etc.) are also not allowed. The connection between the parts of the instruments and the graphical objects representing the geometric objects must be expressed by the use of terms from technical language: "the vertex to the right angle of the square", "the needle of the compass", etc. and from geometric language: "segment", "point", etc.

The instructor is thus led to formulate their intention to act (Petitfour 2017b) without having to use geometric language from the outset, which is not immediately accessible to pupils during their first encounters with geometric concepts. The formulation of action, through the use of technical language, aims, on the one hand, to place the instructor as close as possible within an action situation that they could experience if they were working autonomously, and on the other hand, to foster advancement towards conceptualization. We hypothesize that the use of technical language, since it is structured with the same logical backgrounds as geometric language (Petitfour and Barrier 2019), can be a support to reduce the existing teaching gap between the execution of an instrumented action and its formulation in geometric language.

The constructor handles the instrument according to the instructions given to them: they take the instrument, position it and make the trace as requested (see Figure 8.4).

In Step (2), they must position the instrument by following the instructions carefully, while trying not to meet the instructor's expectations if they have not been explicitly formulated. It is thus a question of acting in "the least likely way", of avoiding any connivance with the instructor, without exhibiting any bad faith. The aim is to fully involve the constructor in the use of geometric knowledge during the construction, and also to give immediate feedback on the quality of the formulations proposed by the instructor, who must then try to improve them by removing the implicit elements. We illustrate this "Least Likely" rule through the previous example of drawing a right angle at A, starting from a segment [AB] (see Figure 8.6).

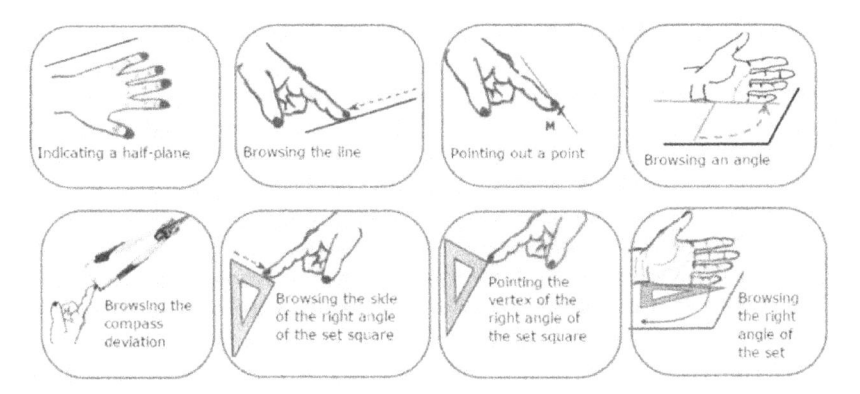

Figure 8.5. *Deictic gestures that can accompany technical language*

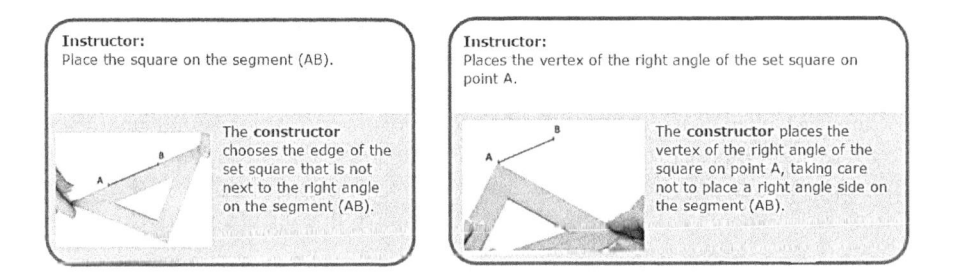

Figure 8.6. *The "Least Likely" rule followed by the constructor*

The dyadic system that we have just presented has the primary beneficial effect of making geometric learning accessible to dyspraxic pupils. Some may find that it does not sufficiently allow for individual work. In order to perform, the device requires the appropriation of technical language, which is not a natural language, and the respect of rules on the parts of both the instructor and the constructor.

We were able to observe over the course of these experiments with sixth grade pupils that the constructor was not always able to follow the "Least Likely" rule, and that sometimes the instructor ended up placing the instrument themselves when they could not express what they wanted to through language. This led us to imagine a dyadic work in the digital environment, using an avatar.

8.4. Designing the human interaction simulator

8.4.1. *General considerations*

Replacing the pupil constructor with an avatar covers several issues. The first is to avoid the biases inherent to human interactions. Since the computer is innately "stupid", in the sense that it is only able to do what is asked of it, the pupil needs to explain things to it as precisely as possible. This would make it think more about its actions. Moreover, the computer does not judge, it does not get impatient and it does not over-interpret the information given. Another challenge is to access the pupil's activity without having observed them in action: the digital tool makes it much easier to keep track of their activity as compared to a pupil–pupil pairing. To encourage pupils to reflect on their actions and to allow them to analyze their choices, the simulators which are known as "part scale" (Béguin and Pastré 2002), that is, those which only reproduce a specific part of reality, are thus the most suitable. We therefore chose to use a human interaction simulator to replace the pupil constructor.

We wanted to use an Embodied Conversational Agent (ECA) (Figure 8.7), that is, a realistic avatar and not a cartoon character. Indeed, the aim is to put the pupil in the same conditions as the dyadic work, and not to let them think that the environment would not respect the physical rules of the real world. In a comic strip, anything can happen if you draw a line along a ruler. ECA is also an interesting choice when compared to peer interaction since it can be much less (or at least differently) emotionally charged, especially for pupils with disabilities (Stowell and Nelson 2007; Brooks 2013). Previous work has shown that it has a positive effect on shyness (Ewing 2006).

To complete the parallelism between the two situations, real and virtual, the position of the avatar and the sound of the exchanges are important. The observer is placed next to the avatar, as in a pupil–pupil dyad, to avoid left-right inversions. The sound of the exchanges is complete, and all the sentences of the avatar, as well as those of the pupil-user, are written and spoken, especially for pupils who would find writing difficult (pupils with visual-spatial disorders, dyslexic pupils). Finally, we wanted to be able to use the simulator on any type of technology: computers, tablets, smartphones, as well as online and offline, to adapt to all the conditions that teachers might encounter in their classrooms. We chose the Virtual Training Suite[5] (VTS) software, which allows us to meet these conditions. We identified several potential benefits of using such software. The scenario can be fully customized and is managed as a graph, divided into scenes. These scenes divide the scenario into drawing tasks.

Figure 8.7. *The Embodied Conversational Agent (ECA) in the simulator*

8.4.2. *Choice of instrumented actions*

For this experimental version, we chose to focus the avatar pupil work on five statements, presented as drawings (Figure 8.8).

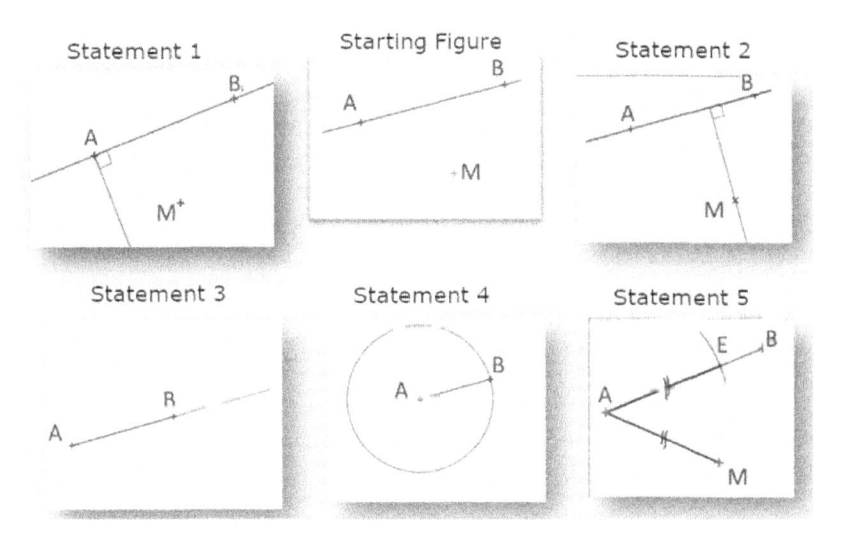

Figure 8.8. *Instrumented actions by the simulator. For a color version of this figure, see www.iste.co.uk/guille/tangible.zip*

---

5 Software from *Serious Factory®*, under educational license granted to l'université de Reims Champagne Ardenne.

Each of these statements corresponds to an instrumented action, that is, an action that leads to the production of a line with an instrument (non-graduated ruler, set square or compass). These five types of instrumented actions have been chosen because they will later encourage a dyadic pupil–pupil to work within the paper–pencil environment, to construct a complex figure through a combination of one or several different types of actions. The five actions are to be carried out from the same starting figure, consisting of three non-aligned points A, B and M, the segment [AB] or the line (AB). These elements are shown in blue in Figure 8.8. A code (in black) marks the perpendicular relations (Statements 1 and 2) and the equality of lengths (Statement 5). The geometric objects to be represented (in red according to Statements 1 through 5) are as follows:

– half-line with origin A perpendicular to the line (AB);

– half-line perpendicular to the line (AB) passing through point M;

– half-line of origin A passing through B (extension of the segment [AB] on the side of point B);

– circle of center A passing through B (or circle of center A, radius AB);

– arc of circle with center A and radius AM intersecting the segment [AB] at point E.

Figure 8.8 is provided to the pupil at the beginning of the assignment, but not the mathematical descriptions above. The pupil is asked to select the instrumented action that they wish to perform (or the one assigned by the teacher according to a pedagogical choice). The following instruction is then given: "We must draw the red line on the starting figure". Note that the software assumes the chronology of the instrumented action by proposing first to choose the instrument, then to position it and finally to trace. This assumption of responsibility by the software only eliminates unrealistic choices, such as tracing before having picked up and positioned an instrument.

8.4.3. *Interaction choices*

Before selecting a statement, the pupil can ask to see the instruments in order to receive information about the specific technical terms necessary for using a compass and a set square (Figure 8.9).

Once a statement has been selected, the first step is to choose which instrument to use: ruler, set square and compass, in order to obtain the red line from the starting figure. The pupil can then see, in video, the hand of the avatar taking the requested instrument.

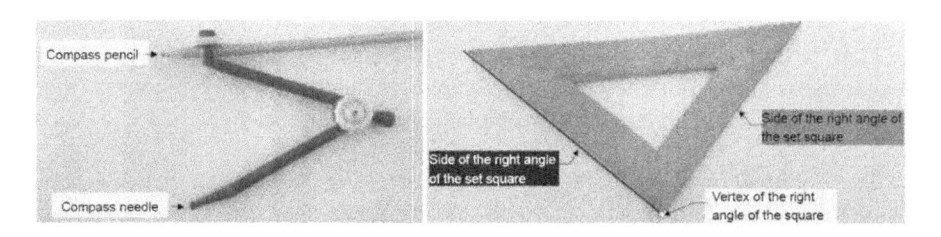

Figure 8.9. *Technical terms*

Types of interactions	Example of Avatar (AV) – Pupil User (PU) interactions			
AV presents the construction to be made: drawing (D), text (T), oral text (OT).	We have to make the red trace on the starting figure.			
PU chooses from a list the instrument to be used: T, OT, image.	Take the set square.		Take the ruler.	Take the compass.
PU confirms their choice.	OK		OK	OK
PU selects from a list the position of the instrument to be performed: T, OT. or AV refuses to act with the chosen instrument: T, TO, video (V).	1. Place the set square on the right (AB). 2. Place a right-angle edge of the set square on the line (AB). 3. On the side of the line (AB) where point M is located, place the right-angle vertex of the set square on point A. 4. On the side of the line (AB) where point M is located, place the right-angle vertex of the set square on the line (AB) with the right-angle vertex at point A. 5. Put back the set square.		1. Place the ruler on point A. 2. Place the ruler horizontally on the line A. 3. Place the ruler vertically on point A and tilt it a little. 4. Place the ruler to make a right angle with (AB). 5. Put the ruler back down.	AV: I don't see how I can do this with the compass.
PU confirms their choice.	OK		OK	AV returns to the instrument menu.
AV performs the action: V	Video		Video	
AV verifies that the positioning of the instrument is what the pupil wanted: T, OT, image.	Do I trace?		Do I trace?	
PU answers: T	Yes	No: actions menu	Yes	No: actions menu
AV performs the action: V	Video		Video	
AV asks if the trace is correct: T, OT.	Do you think the trace is correct?		Do you think the trace is correct?	
PU answers: T	Yes	No: actions menu	Yes	No: actions menu
AV compares the expected and actual traces: D, T.	Was it successful?		Was it successful?	
PU responds: T	Yes	No	Yes	No
AV formulates the trace in geometric language: T, OT, D.	We have drawn the line...	Back to actions menu	No: back to previous menu	Back to actions menu

Figure 8.10. *Interactions with the human interaction simulator. For a color version of this figure, see www.iste.co.uk/guille/tangible.zip*

Four or five positions of the chosen instrument in relation to the geometric objects present (see columns 2 and 3 in Figure 8.10) are then what is proposed (Step 2). One is valid, and the others are incomplete or not precise enough.

The formulations were chosen on the basis of those proposed by pupil instructors, during experiments with dyad work in the paper-and-pencil environment, for these same types of tasks. The pupil selects the instruction that seems to be appropriate, and then watches a video of the avatar's hands performing the requested positioning, while observing the "Least Likely" rule (Figure 8.6).

The pupil can go back on their decision or confirm the avatar to carry out the trace. In the latter case (Step 3), they see a video of the avatar's hands making a trace from the position of the chosen instrument.

At different moments throughout the interaction, the pupil has the opportunity to confirm or to go back on their choices, based on the effects presented to them in the video. We summarize these in Figure 8.10, by taking as our example Statement 1.

8.4.4. *Ergonomic considerations*

Figure 8.10 also helps to understand our ergonomic choices. They take into account three types of difficulties that the pupil may encounter: decentration (Radford et al. 2008), that is, the ability to imagine another individual's point of view; reading difficulty, as the comorbidity of dyspraxia is often associated with dyslexia or dysorthographia and inhibition (Friso-van den Bos et al. 2013), that is, the ability to inhibit, to hold back the most obvious answer and to wonder if a more appropriate solution might not exist.

Decentration: the avatar is positioned to simulate a side-by-side configuration. The videos are filmed in the first person (through the eyes of the constructor) so that the pupil is either in the avatar's shoes or looking over the avatar's shoulder, as is the case in pupil–pupil dyad work.

Reading difficulty: the pupil interacts with the artifact through a list of written choices, but may also request oralization. Vandenbroucke and Tricot (2018) highlighted that the dyslexic pupil did not have better comprehension of oralized texts when compared to texts they would read in written form. It is the time factor available for reading that is decisive. We leave the choice to the pupil to read and/or listen to the sentences without a time limit. The texts, when they refer to an instrument, are also illustrated through pictures. The objective is that reading is not a limiting factor to the pupil's success.

Inhibition: in order to encourage inhibition and reflection, pupils are systematically asked to confirm their choices, whether it is the instrument or the instruction given to the avatar. Once the instrument is placed, the pupils have to confirm that it is what they expected, and when the trace is made, they have to confirm that the goal has been reached. This absence of automatic validation by the software, concretized by the question "Do you think that the line is correct?" makes it possible to verify that the goal to be reached is well identified by the pupil, which makes it possible to analyze the source of the difficulty: related to the trace or to the comprehension and identification of the construction to be carried out. All these interactions also make it more costly to systematically try out all the proposed choices, when compared to anticipating and finding the correct solution.

To facilitate the teacher's work, we have chosen to produce a summary of the activities for each pupil and not an exhaustive recording of each of the actions chosen, which would be difficult to use. Therefore, the software memorizes nine synthetic variables: the number of correct choices (N1), that is, choosing the correct formulation and instrument, and the number of incorrect choices (N2); the number of times the pupil correctly anticipated that the positioning of the instrument was valid (N3), and when the positioning of the instrument was invalid (N4); the number of times the pupil did not correctly anticipate that the positioning of the instrument was valid (N5), or that it was invalid (N6); the number of times the pupil correctly identified that they had reached (N7), or not (N8), the goal and the number of times the pupil did not identify that they had reached the goal (N9).

8.5. Initial experimental results

We carried out a first experiment articulating a work between the two types of dyads, pupil–pupil and pupil–avatar, in a class of sixth graders that included Jim, a pupil with special educational needs which we will specify in greater detail later on. At the same time, we tested the pupil–avatar dyad working arrangement in a fourth-grade class (9–10 years old). Starting with the fact that a pupil is considered as having a disability when they are more than 18 months behind the usual learning for their age group, we chose, for this experiment, to collect information from pupils in the sixth grade, Jim's age group, and pupils in the fourth grade, who are therefore about two years younger than Jim. The fourth-grade class served as a control class for the software performance, and only the session with the avatar was proposed. We present the data collected, as well as our observations and analyses for these first tests.

8.5.1. *Data collected*

In the fourth-grade class in 2018, 11 pupils worked for one session on the simulator in pupil–avatar dyads. No pupils were diagnosed with disabilities. The pupils could freely choose the constructions to be made from among the five proposed constructions (Figure 8.8). No specific work, either in pupil–pupil dyads or on the construction instruments, was done beforehand. The number of constructions was not limited, but the session lasted 20 minutes. The session was filmed in its entirety, and the results from the software were recorded. By combining these two data collections, it is possible to track the pupils' choices on the software and to summarize them (Figure 8.15). In the sixth-grade class, we filmed two one-and-a-half hour sessions, conducted one week apart during the 2018–2019 school year. The first session began with focusing the whole class on the functions of the set square and the compass, as well as on the vocabulary related to the components of these instruments in terms of their relations to geometric objects. The pupils then each worked for about 15 minutes with the software (pupil–instructor and avatar–constructor dyad) on the five constructions shown in Figure 8.8. During this session, we filmed Jim, and then collected his work and the work on the simulator of six other pupils in the class. In the second session, pupils worked in the paper-and-pencil environment, first on the same five constructions (pupil–pupil dyads), alternating roles as instructor and constructor. They then worked on the construction of a square ABCD with center O, to be made from a given segment [AO], in pupil–instructor and pupil–constructor dyads; an extra observer pupil was added to the dynamic.

The latter was responsible for ensuring that the rules established for the roles of instructor and constructor were followed, and for collecting the tracing program developed by the dyad. In addition, before starting the construction, the three students had to agree on the construction technique they were considering. The work support consisted of the written instruction: "Construct the square ABCD with center O using the non-graduated ruler, the set square and the compass" and the drawing of a segment [AO]. Prior to the two sessions described, we gave Jim different geometric construction tasks as a diagnostic assessment to collect filmed data on what he could do on his own. In the following section, we review this assessment after disclosing a few key elements of his occupational therapy assessment.

8.5.2. *Jim's diagnostic evaluation*

Jim, 11 years old, has difficulties with written language (dyslexia and dysgraphia). His psychomotor assessment shows, among other things, immature spatial organization and difficulties with finger dexterity and oculomotor coordination. His cognitive evaluation shows good verbal skills, speedy information processing, correct perceptual reasoning, but insufficient working memory. We have observed, during the realization of geometrical constructions with Jim's instruments, manipulative difficulties (his hold of the set square and the compass is not adequate enough for successful tracings), as well as organizational difficulties (e.g. Jim spends a lot of time in the preparing his compass, playing around with the lead that he does not succeed in fixing correctly). At the language level, he uses the geometric terms "half-line", "line", "segment" and "circle" correctly, but does not formulate the characteristic elements: he speaks of the "line M" to name the "line perpendicular to the line (AB) passing through M" and of "circles" without providing any precision. The manipulative and organizational difficulties, the geometric language not yet acquired and Jim's good verbal skills already suggest the potential benefits a dyadic work arrangement may provide him. We were also able to observe, as with other pupils in the class, the implementation of erroneous construction techniques, which did not result in obtaining correct figures. Jim, in fact, does not seem to distinguish between the geometric relations perceptually obtained when using the appropriate versus those obtained using the inadequate instrument. He relies on his global perception of the figures, considering only the length of the sides.

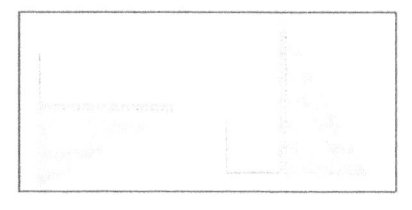

(a) Drawing a square　　　　　(b) Extending a segment

Figure 8.11. *Positioning of instruments by Jim*

For example, in order to trace a square (Figure 8.11a), he uses his set square for its function of "drawing a straight line of a given length": no side of the right angle of the set square rests on an already drawn side of the square, and the scale "0" is used instead of the vertex of the set square's right-angle being positioned at the end of the already drawn side. To extend a side (Figure 8.11b), he places the set square perceptually, without supporting the edge. Jim, thus, does not distinguish between "appearing to be aligned" (perceptual control) and "needing to be aligned" (instrumental control) when extending a segment. His construction techniques are wrong, and he is not aware of it. In fact, according to him, there are right angles because he used the set square, and he believes that his production is correct. A validation with the tracing paper would probably lead him to question the precision of his tracings more than the correctness. In terms of instrumental genesis, he recognizes the set square as an artifact for constructing right angles, but the associated instrumental schemes are not adapted. Let us now look at the feedback that dyadic work has provided with respect to these erroneous construction techniques concerning the extension of a segment and the drawing of a right angle.

8.5.3. *Analysis of the first experimentation*

8.5.3.1. *Extending a segment*

Concerning Statement 3 (Figure 8.8), the choices proposed by Jim for the positioning of the ruler by the avatar, with the aim of extending the segment [AB], show that he does not take into account in an instrumental way the alignment of the segment [AB] with the extension to be traced. He chooses successively the following three proposals, which he abandons (selected answer: "No, I will choose something else") when he sees the positioning done by the avatar: "Place the ruler in the alignment of A and B, starting from B", "Place the ruler on point B" and "Place the ruler straight from point B". At the third proposition, he exclaims "Ah, she doesn't understand anything!". Finally, he chooses the proposition "Place the ruler on the segment [AB]" and validates it.

The instruction to place the ruler "from point B" does not allow the edge of the ruler to simultaneously be on the segment [AB], and beyond point B, in order to accommodate the extension. The indications "in line with A and B" and "straight edge" suggest a perceptive consideration of this alignment. The time spent reading the proposals and moving from one to the other, materialized with the pointer by highlighting the text, leads us to believe that Jim did not make random choices and that his choices correspond to formulations for the positioning of the ruler that he considered to be valid. This interpretation is supported by the fact that we were able to observe Jim proceeding in the same way, in his constructions carried out autonomously (see section 8.5.2). The feedbacks proposed by the avatar produce the expected effects: they lead Jim to a valid instruction.

A week later in dyad work in the paper-and-pencil environment for this same construction, Jim formulates the following instruction to Andy:

> Take the ruler. Place it straight on the line AB, and it must pass through A and B.

Andy asks him to repeat himself, and Jim then rephrases it like this:

> "You have to place the ruler on the point A so that it passes through B, it has to pass through B.

This wording does a good job of positioning the ruler correctly. Andy places it as expected. He extends the segment [AB] on either side of A and B, responding to Jim's request to trace. Andy then points out that the extension should have been only on the side of point B.

In the next construction of the square ABCD, with center O, from segment [AO], Jim decides to be a constructor (despite his manipulative difficulties). Andy is an instructor, Matthias an observer. We relate the episode concerning the extension of the segment [AO] desired by the instructor Andy, after the construction of two circles of radius AO, one of center A and the other with center O (Figure 8.12).

Andy asks Jim to take the ruler, but Jim insists on taking the set square.

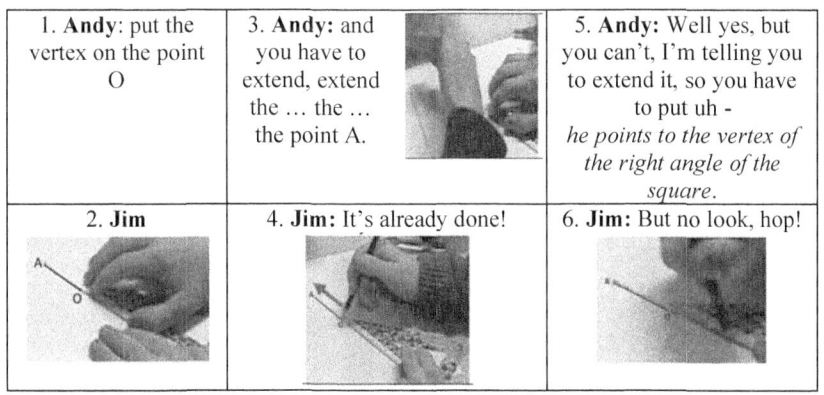

1. **Andy**: put the vertex on the point O	3. **Andy:** and you have to extend, extend the … the … the point A.	5. **Andy:** Well yes, but you can't, I'm telling you to extend it, so you have to put uh - *he points to the vertex of the right angle of the square.*
2. **Jim**	4. **Jim:** It's already done!	6. **Jim:** But no look, hop!

7. **Andy**: To extend you can …
8. **Jim** *interrupts him*: There you go!
9. **Andy:** Well yes, but uh point A not point O
10. **Jim:** There you go

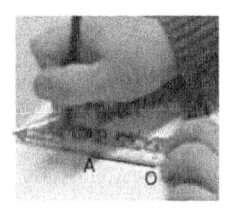

11. **Andy:** And do the same for point O
12. **Jim**: I already did
13. **Andy:** Try to do it more precisely
14. **Jim:** I already did!
15. **Andy:** No, that's not accurate! He points
16. **Jim:** Yes, it is!

Figure 8.12. *Positioning of instruments by Jim*

In this episode, Andy expresses what he is aiming at, the extension of the segment [AO] on the side of point A, with the linguistic instruction (3) "you have to extend", completed by a deictic gesture. The formulation that he struggles to find and that he uses "extend the ... the ... the, the point A" is incorrect, and the positioning of the square requested in 1 – "the vertex on the point O" – is not sufficient. The positioning of the set square proposed by Jim's action (2) forces Andy to be more explicit, just as it would have done with the avatar. Andy, however, does not seem to be able to improve his wording by himself. Perhaps the choices offered by the avatar would help him in this situation. Note that Jim does not let Andy pick the instrument of his choice, which would not be the case with the avatar. Nor does he encourage him to specify the positioning of the instrument (line 8, Jim cuts Andy off). However, when Jim places the square with "the vertex on point O", he does put it in alignment with the segment [AO] with the intention of making an extension of [AO] on the side of point O (action 4). For a similar instruction: "Place the ruler on point B", the avatar proposes a positioning of the ruler that leaves no doubt that it is not appropriate (Figure 8.13).

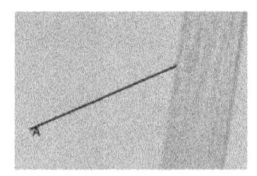

Figure 8.13. *Positioning the ruler in response to: "Place the ruler on point B"*

Jim is not able to produce such feedback since he did not perceive that the instruction given by Andy was incomplete. The positioning of the square that Jim proposes corresponds to the formulations that he had chosen with the avatar for construction No. 3: "Place the ruler in alignment with A and B, starting from B", "Place the ruler on point B", "Place the straight edge of the ruler from point B" and that he therefore still considers acceptable. Andy's requests for clarification (line 13 and line 16) are not sufficient for Jim to perceive that his construction is not correct.

8.5.3.2. Tracing a right angle

For Statement 1 (Figure 8.8), Jim chooses the instruction "Place one side of the set square on the line (AB)". The positioning of the square proposed by the avatar (Figure 8.14) leads him to choose the correct instruction. He then directly gives the correct instruction for Statement 2 (Figure 8.8).

As the instructor for this same construction a week later, Jim asks Andy to take the square and place the side of the right angle on the line (AB), again leaving the second constraint implicit. Andy performs this positioning, taking care that the other side of the right-angle does not pass through point M. Jim then asks Andy to "move the set square back", to which Andy responds, mimicking the avatar, "I don't see how to do that."

Figure 8.14. *Statement 1 and feedback from the avatar*

Jim tries new formulations, such as "You have to put the side of the right-angle on point M and it has to pass through line AB", which do not result in the desired positioning, because Andy is playing the game of not decoding the implicit. Jim gets a little annoyed. He ends up positioning the set square himself, which would be impossible with the avatar. Andy then asks if he can trace, then congratulates Jim, imitating the avatar. In this construction, Jim knew how to set the square correctly, but he could not formulate it precisely. When Jim, Andy and Matthias later try to trace the second side of the set square with side [AO], Jim will propose a length transfer in a guessed direction, followed by a check with the set square. Jim is not yet able to find a valid usage of the set square to construct a right angle, for a construction where two properties (the length and the right angle) are taken into account.

8.5.3.3. *Summary of the results obtained in the simulator*

Our goal is to see if Jim's scores are comparable to those of his peers, and thus determine if his disability singles him out, or if they are comparable to those of pupils 18 months younger, and thus close to the skills assumed to be related to his disability. We have compiled the pupils' scores in Figure 8.15.

The last row (N9) is for pupils who did not identify that they got the correct drawing. Nine pupils made this type of error, including Jim. These pupils had trouble identifying the goal, even when it is superimposed on the drawing (indeed, the software superimposes the drawing to be made on top of the one the pupil made before asking if they think they got it right), and it is clear that they had trouble choosing the correct actions to perform.

	6th Grade Pupils						4th Grade Pupils										Jim
	1	2	3	4	5	6	1	2	3	4	5	6	7	8	9	10	11	
N1 (%)	30	47	42	100	47	29	38	45	38	33	36	46	52	50	38	21	60	39
N2 (%)	70	53	58	0	53	71	63	55	62	67	64	54	48	50	62	79	40	61
N4	1	0	1	1	0	2	0	0	0	2	4	2	0	0	0	2	0	1
N6	2	1	0	1	1	0	3	1	2	1	2	2	1	0	2	5	1	1
N3	3	4	3	0	0	2	0	3	2	1	2	1	2	1	4	1	1	4
N5	0	0	0	1	0	0	3	0	0	0	0	0	0	0	0	0	1	0
N7	2	1	0	1	1	0	4	1	2	1	2	2	1	0	2	5	0	1
N8	0	0	0	1	0	0	3	0	0	0	1	0	1	0	1	2	1	1
N9	1	0	1	1	0	2	0	0	0	2	4	2	0	0	0	2	0	1

N1 (in %): correct choices
N2 (in %): incorrect choices
N3: anticipation of validity
N4: anticipation of non-validity
N5: non-anticipation of validity
N6: non-anticipation of non-validity
N7: identification of obtaining goal
N8: identification of not obtaining goal
N9: non-identification of obtaining goal

Figure 8.15. *Software result summary*

Line 6 (N5) corresponds to pupils who did not see that the drawing was correct and Line 4 (N6) corresponds to pupils who did not see that the drawing was incorrect. In both cases, it appears that it is difficult for these pupils to visualize the objective. This is the case for Jim. The first two rows give the percentages of correct and incorrect choices. It is reasonable to assume that the pupils who made many incorrect choices, but showed good anticipation, were using the trial-and-error principle to some extent. They may have spotted the correct design but acted too quickly, or, they established a relationship with the software that encouraged them to make the correct choices. This seems to have been the case for the sixth grader, No. 3, and the fourth grader, No. 4, for example. Finally, 6th grader No. 4 attempted too few actions (three in all) for the analyses to be relevant. Regarding Jim specifically, we note that, although his scores are low, he is indistinguishable from other sixth graders who are not dyspraxic, and his scores are higher than several fourth graders.

Even though it is only an exploratory study on a few pupils, these first results suggest that the software allows the dyspraxic pupil to work like other pupils.

8.5.4. Conclusion

The objective of this study was to design an artifact that could help pupils with dyspraxia to do geometry. This exploratory work aims at establishing a proof of concept. The software we produced is part of a more global approach, which aims to bring dyspraxic pupils to use a technical language (Petitfour 2017b) allowing them to produce instrumented actions, without having to resort to geometric language when they have not yet mastered it. This language is used to communicate with a virtual avatar or with real pupils who are able to interpret it and put it into practice.

How can pupils be encouraged to use the appropriate language? Pupils must manipulate a form of language that is not natural to them, and are presented with a list of technical language choices (ibid.). It should be noted that we have not explored other options, such as artificial intelligence tools to interpret pupils articulated natural language (Collobert et al. 2011), along the lines of what commercial voice assistants are capable of doing. The idea would be to let the pupil formulate what they want the avatar to do, "in their own words", and to identify which of the avatar's actions this corresponds to. For example, if the speech segment "the side of the right-angle (of the set square) on the line" is not detected, the avatar places "the wrong side of the set square" along the line.

Does using language as a substitute for manipulation create a new challenge? A first consideration is the decision to allow pupils to choose from a finite list, reflecting on their choices, and change their selection. They can choose a formulation, see the result, and change their minds. However, it is possible that the reduction of the universe of possibilities induces a catachresis type of instrumentalization (Rabardel 1995); the tool is bypassed to find the right answer, all choices are tested and the longest sentence (which is very often the right one) is selected.

In terms of the potential benefits of alternating pupil–pupil and pupil–avatar dyads, our initial results indicate that alternating the two approaches could be beneficial for all pupils. Our initial observations show that even pupils who were not diagnosed as having special educational needs performed similarly to or worse than Jim. The benefits of such work therefore go beyond finding a way to compensate for a disability and could be a new tool for general learning. Pupils built a relationship with the artifact, which included helping them understand the role of the constructor. Pupils took the avatar's responses, and some actions, as a model for their own behavior. This last observation leads us to new hypotheses regarding the dialectic between pupil–pupil and pupil–avatar work. This scale of analysis opens a perspective in which dyads can be seen as a duo of digital and material artifacts (Maschietto and Soury-Lavergne 2013; Voltolini 2018). In this case, mutual contributions and complementarities could be considered. This approach needs to be developed.

Finally, a challenge that should be addressed in future research is to propose a more complete set of instrumented actions. Furthermore, working as closely as possible with a class over time will allow the development of new alternatives for teaching and learning geometry, not only for pupils with special educational needs, such as those with dyspraxia, but also, more generally, for all pupils.

8.6. References

Béguin, P. and Pastré, P. (2002). Working, learning and designing through simulation. In *Proceedings of the 11th European Conference on Cognitive Ergonomics*, Bagnara, S., Pozzi, S., Rizzo, A., Wright, P. (eds). Istituto de Scienze e Tecnologie della Cognizione, Padua.

Berthelot, R. and Salin, M.H. (1992). L'enseignement de l'espace et de la géométrie dans la scolarité obligatoire. Thesis, Université Sciences et Technologies-Bordeaux I.

Brooks, E. (2013). Ludic engagement designs: Creating spaces for playful learning. In *Universal Access in Human–Computer Interaction: Design Methods, Tools, and Interaction Techniques for eInclusion*, Stephanidis, C. and Antona, M. (eds). Springer-Verlag, Berlin.

Collobert, R., Weston, J., Bottou, L., Karlen, M., Kavukcuoglu, K., Kuksa, P. (2011). Natural language processing (almost) from scratch. *The Journal of Machine Learning Research*, 12, 2493–2537.

Elbasan, B. and Kayihan, H. (2012). Motor performance and activities of daily living in children with developmental coordination disorder. *Journal of Novel Physiotherapies*, 2(2) (#1000107).

Ewing, A. (2006). Increasing classroom engagement through the use of technology [Online]. Available at: http://www.mcli.dist.maricopa.edu/mil/fcontent/2005–2006/ewing rpt.pdf.

Friso-van den Bos, I., van der Ven, S., Kroesbergen, E., van Luit, J. (2013). Working memory and mathematics in primary school children: A meta-analysis. *Educational Research Review*, 10, 29–44.

Houdement, C. and Kuzniak, A. (2006). Paradigmes géométriques et enseignement de la géométrie. *Annales de didactique et de sciences cognitives*, 11, 175–193.

INSERM (2019). *Trouble développemental de la coordination ou dyspraxie*. Éditions EDP Sciences, France.

Maschietto, M. and Soury-Lavergne, S. (2013). Designing a duo of material and digital artifacts: The pascaline and Cabri Elem e-books in primary school mathematics. *ZDM, The International Journal on Mathematics Education*, 45(7), 959–971.

MEN (2018). Programmes du cycle 3, en vigueur à compter de la rentrée de l'année scolaire 2018–2019. BOEN no. 30 du 26 juillet 2018. Ministère de l'éducation nationale, France.

Petitfour, E. (2015). Enseignement de la géométrie à des élèves dyspraxiques visuospatiaux inclus en classe ordinaire. *Recherches en éducation*, 23, 82–94.

Petitfour, E. (2017a). Enseignement de la géométrie en fin de cycle 3. Proposition d'un dispositif de travail en dyade. *Petit x*, 103, 5–31.

Petitfour, E. (2017b). Outils théoriques d'analyse de l'action instrumentée, au service de l'étude de difficultés d'élèves dyspraxiques en géométrie. *Recherches en didactique des mathématiques*, 37(3–2), 247–288.

Petitfour, E. (2018). Quel accompagnement en géométrie pour des élèves dyspraxiques ? *Grand N*, 101, 45–70.

Petitfour, E. and Barrier, T. (2019). D'un cadre d'analyse de l'action instrumentée en géométrie à l'élaboration d'un dispositif de travail en dyade au cycle 3. In *Actes de la 19e école d'été de didactique des mathématiques ARDM*, Coppé, S. and Roditi, E. (eds). La Pensée Sauvage, Grenoble.

Rabardel, P. (1995). *Les hommes et les technologies : approche cognitive des instruments contemporains.* Armand Colin, Paris.

Radford, L., Schubring, G., Seeger, F. (2008). *Semiotics in Mathematics Education: Epistemology, History, Classroom, and Culture*. Sense Publishers, Rotterdam.

Sagot, J. (2005). TGT : un outil pour construire la géométrie ? *Réadaptation*, 522, 45–48.

Sarama, J. and Clements, D. (2016). Physical and virtual manipulatives: What is "concrete"? In *International Perspectives on Teaching and Learning Mathematics with Virtual Manipulatives*, Moyer-Packenham, P. (ed.), Springer, Cham.

Stowell, J. and Nelson, J. (2007). Benefits of electronic audience response systems on pupil participation, learning, and emotion. *Teaching of Psychology*, 34(4), 253–258.

Straesser, R. (2002). Cabri-géomètre: Does dynamic geometry software (DGS) change geometry and its teaching and learning? *International Journal of Computers for Mathematical Learning*, 6(3), 319–333.

Vaivre-Douret, L., Lalanne, C., Ingster-Moati, I., Boddaert, N., Cabrol, D., Dufier, J.-L., Golse, B., Falissard, B. (2011). Subtypes of developmental coordination disorder: Research on their nature and etiology. *Developmental Neuropsychology*, 36(5), 614–643.

Vandenbroucke, G. and Tricot, A. (2018). La présentation orale de textes narratifs améliore-t-elle la compréhension d'élèves dyslexiques de CM2 ? *Analyse neuropsychologique des apprentissages chez l'enfant*, 152, 111–121.

Voltolini, A. (2018). Duo of digital and material artefacts dedicated to the learning of geometry at primary school. In *Uses of Technology in Primary and Secondary Mathematics Education: Tools, Topics and Trends*, Ball, L., Drijvers, P., Ladel, S., Siller, H.-S., Tabach, M., Vale, C. (eds). Springer, Cham.

9

Research and Production of a Resource for Geometric Learning in First and Second Grade

9.1. Presentation of the ERMEL team's research on spatial and geometric learning from preschool to second grade

The research of the ERMEL team (team associated with the French Institute of Education of the École Normale Supérieure de Lyon) on spatial learning and geometric learning from kindergarten to third grade (5–8 years old) has two goals.

On the one hand, we aim to analyze teaching challenges by taking into account the needs of teachers, identifying the components of the targeted learning and specifying the knowledge that the pupils initially have at the beginning. We therefore have to clarify hypotheses about learning, to construct and experiment with situations that guarantee its acquisition and to develop progressions that take into account the entire school program. The results of these experiments, carried out in numerous classes over several years, have led us to question the choices that were the basis of certain learning processes, and to modify these systems.

On the other hand, our goal is to produce resources (Douaire et al. 2020) for teachers and trainers, offering them a coherent vision of spatial and geometric learning in terms of both the content covered and the relationship between pupils and mathematics, as well as the set of tasks that the teacher must conduct.

Chapter written by Jacques DOUAIRE, Fabien EMPRIN and Henri-Claude ARGAUD.

9.1.1. *Origins of the research*

This research has several origins. First of all, teachers frequently state that they do not have enough problems that pupils can apply themselves to, and that the activities are somewhat repetitive from one year to the next, from the beginning of primary school onwards. Moreover, we perceived a strong difference with teaching challenges in the digital domain. Indeed, the reflection on the learning proposed for this age group, as much as for the acquisition of the number and numeration as for the calculation or resolution of additive problems, can be based on numerous works in mathematics education. However, for geometric learning at these levels, analogous analyses that take into account the different aspects of a concept, or the pupils' level of knowledge, are rarer.

In a previous research study (ERMEL 2006), we developed a progression that privileged, for third and fourth grades (8–10 years old), an entry into the learning of relations (alignment, parallelism, perpendicularity, distance, symmetry), then developed a summary on the properties of objects in the fifth grade (10 and 11 years old). This approach proposed the resolution of spatial and geometrical problems in meso-space, micro-space (Berthelot and Salin 1992), and particularly in graphical space, and associated with the use of dynamic geometry software (DGS), which can simulate the conditions of other types of space such as meso-space, for example. This choice of teaching method also allowed teachers to move away from a discourse on geometric figures, sometimes limited to their description or to the statement of their properties, in order to privilege the resolution of problems which allowed teachers to identify the knowledge that the pupils were really able to put into practice. This structuring around relations gave an overall coherence to learning in order to explain, for example, the relations between the objects of the plane and those of the space, the role of problems in this learning, the importance of the group sharing stages, etc., a coherence that we have, to a certain extent, preserved for our research on geometric learning from kindergarten through to second grade.

9.1.2. *Introduction to the chapter*

We present here the evolution of research questions concerning the drawing of straight lines and the discovery of the properties of plane figures and solids, by carrying out, each time, an analysis of the knowledge of pupils in the first and second grade (6–8 years old) as well as an analysis of the problems being proposed. We also address questions concerning the implementation of situations and the progression and structuring of the resource.

9.2. Learning to trace straight lines

In frequent practices, straight line traces are introduced as tools to reproduce geometric drawings; properties of the straight line are not constructed through problematized experiences in meso-space or on traces, but often through the evocation of prototypical objects (solids, figures). These practices do not always allow for the differing initial conceptions of pupils to be expressed, whether these are erroneous or potentially useful. We propose to take these conceptions into account in order to establish a teaching method that challenges them, if necessary.

9.2.1. *Significance of the straight line*

Among the various spatial meanings of the straight line advanced in primary school, we wanted, initially, to specify those upon which we could rely on to promote an evolution in knowledge.

A straight line can represent a material object, deformable or not (a stretched wire, the edge of a table, a ray of light, a fold in a sheet of paper, etc.), or the boundary between two areas (the edge of a solid, the side of a figure). It can be used to determine a hidden object (by means of sighting, for example) as well as a set of points aligned with other points or located at the same distance from the points of a straight line. It can be produced by tracing straight lines, in graphical space with the ruler or with geometry software. Should all of these meanings be apprehended at the beginning of primary school or should some be favored, and if so, which ones? In order to answer this question, and to clarify pupils' knowledge, we have chosen, as a first step, to propose alignment problems in the meso-space before modeling them using pencil and paper.

9.2.2. *Initial hypotheses*

We had also noted, during the previous research project (ERMEL 2006), that some pupils at the beginning of the third grade were surprised to note that a line could be drawn between two distant points on a sheet of paper. Our first hypothesis was therefore to envisage, at the beginning of the current research, an apprehension of the notion of alignment, favored by the passage of spatial experiments centered around the idea of hiding a distant object by another one that is closer to the observer (solved by means of sighting, the use of taut wires, or tracings on the ground). These activities proposed in the meso-space were then transferred to paper.

The aim was to highlight the necessity of using straight line to produce the solution. The proposed problem (Figure 9.1) was employed to determine, in the space of the playground, a location wherein, for example, a red cone hides the green cone, and the yellow cone hides the orange cone. However, we found that when this problem was then presented on an A3 sheet of paper, many pupils thought they had succeeded in finding a location, even though they had only drawn broken straight lines. The invalidation of their solution, for example, with a large ruler, was a real surprise for them. Moreover, the procedures previously developed by the pupils when they are themselves part of the setup in the meso-space (producing or checking, notably by aiming, an alignment of cones in the playground), are no longer applicable for solving problems with pencil and paper. In conclusion, the modeling on paper of these activities experienced in space does not guarantee the complete understanding of the properties of a straight line.

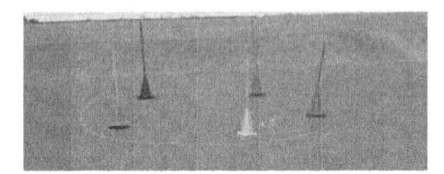

Figure 9.1. *The cones. For a color version of this figure, see www.iste.co.uk/guille/tangible.zip*

A second hypothesis dealt with the effect of using dynamic geometry software as an intermediary between solution-finding using meso-space and paper. Continuing the situation using the Cabri Géomètre™ software made its resolution using a microcomputer possible. The pupils had easy access to the software's tools, allowing them to produce a correct solution by intersecting two straight lines, especially favoring aerial views. However, subsequent experiments did not allow us to verify an improvement of resolution procedures, in particular for the cones, when the activities previously conducted on microcomputer were again proposed with pencil and paper.

A third hypothesis was therefore to privilege the learning of straight lines and the discovery of its properties (the straightness) of the straight line, independent of its use as a means to solve a problem involving the alignment of points. It was hence a question of making explicit the criteria for the implementation of adequate traces, as well as the constraints on the use of the chosen instrument. We briefly describe two situations developed and experimented with this aim.

9.2.3. *The RAYURE situation*

From the first grade onwards, in order for the pupils to understand that straight lines can be extended, and to learn how to do so, it seemed necessary to us that they should be able to draw straight lines in order to solve problems (Figure 9.2), and thus to control and improve them, leading them to analyze their errors and to make explicit the criteria for judgment. The problem is presented on an A3 sheet of paper, and the pupils do not have an instrument longer than 20 cm. They are thus led to moving the ruler in order to extend straight lines, the instruction being: "You must extend the traced lines, and not add any new ones."

Figure 9.2. *The student worksheet*

Different types of productions in the first grade, during the second quarter, with pupils who have already had various opportunities in school to draw straight lines with a ruler, are shown in Figure 9.3. In particular, we observe:

– freehand traces (Figure 9.3), the paradox being that the pupil is holding a ruler in their hand;

– tracings made partially with a ruler but sometimes extended by hand (Figure 9.3(b)); some of these also show broken lines;

– correct tracings with possible inaccuracies in the continuity of the lines (Figure 9.3(c)).

On their own, some pupils use a large ruler to check the straightness of their trace (Figure 9.3(d)).

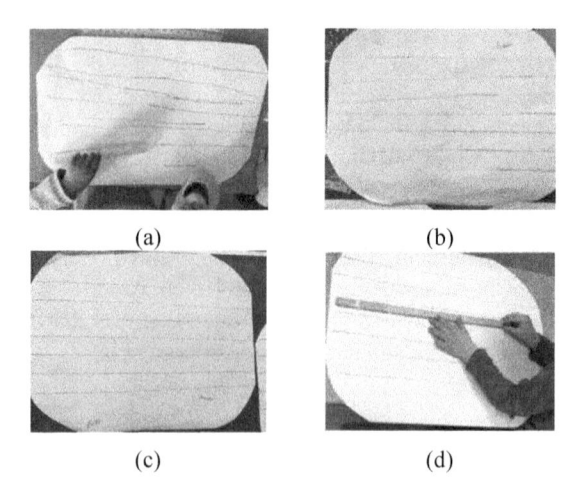

Figure 9.3. *(a) Freehand traces; (b) mixed traces; (c) straight line traces; and (d) subsequent check*

9.2.4. *Using straight lines*

Other activities aim to show how straight lines can be used to represent the relative position of parts. Two situations are proposed, one in the first grade where the pupils have to represent a stack of rectangles (Figure 9.4(a)) by completing a drawing of which they only have the outline (Figure 9.4(b)).

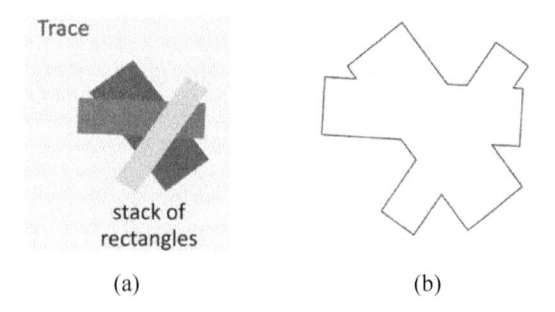

Figure 9.4. *(a) Stacked rectangles and (b) trace to be completed. For a color version of this figure, see www.iste.co.uk/guille/tangible.zip*

Another situation is proposed in the second grade, where pupils have to determine whether a rectangular strip (shaded in Figure 9.5), cut into two pieces and whose junction is hidden by a paper cover, has been repositioned correctly or not:

for example, has one of the pieces been slid sideways? Several problems are submitted to them: correct positioning, slight translation or rotation of one of the two parts. These possible displacements are undecidable perceptually and therefore imply the use of straight lines to extend the sides.

Figure 9.5. *Broken rectangle*

9.2.5. *A few summary elements*

During the group discussions, the analysis of the productions and the formulation by the pupils on the conformity and differences between the tracings make it possible to make various criteria of judgment explicit, which constitutes one of the components for this learning. These criteria relate in particular to the understanding of the task, the technical mastery of the ruler and the technology involved in making a transfer with the ruler.

Indeed, the pupil must understand the goal to be reached: for example, that a straight line must be extended, which contributes to the establishment of the properties of a straight line as early as first grade – the use of the straight line to solve problems involving the alignment of points typically approached in the third grade.

The mastery of the tracing technique (positioning the ruler, keeping it pressed in the middle without letting the fingers protrude, placing the pencil, etc.) must be practiced during various sessions, including outside of mathematics.

The techniques (transferring the ruler to place an intermediate point or sliding the ruler over a sufficient part of the line) must be formulated so as to be shared and understood.

9.3. Plane and solid figures

9.3.1. *Findings and assumptions*

As of kindergarten, pupils become familiar with objects in space and on the plane, the recognition of which is, for the most part, based on the visual perception of their overall shape, with, as a result, possible confusion between neighboring objects: for example, a rectangle whose length and width are close is often qualified as a square by pupils at the beginning of primary school.

In order for knowledge on the properties of figures and solids to not be lessened to their respective statements, sometimes formulated on the teacher's orders, the properties must appear to pupils as necessity and not convention. Without going back over the history of our research, let us quote four of our main successive hypotheses.

The first hypothesis, which is not specific to us, is that the study of solids and that of plane figures are in interaction: the recognition of solids leads to the identification of their faces, and the analysis of plane figures allows in return to deepen the knowledge of solids.

The second hypothesis is that, in order to overcome the ambiguities of perception, especially visual perception, it is preferable to have pupils identify the properties necessary to distinguish perceptually very close objects in the plane or in space by solving comparison problems, rather than to constitute a priori a classification around objects that are perceptually different. The evolution of these criteria of judgment being, here also, a learning goal.

The third hypothesis is that the gestures with objects in space or in the plane: "sliding", "turning", "flipping", etc., implemented by pupils since kindergarten, and common to the resolution of various problems, are finalized actions whose success or failure the pupil can observe. These carry geometric and spatial relationships, constituting procedures that can be made explicit, and whose effects can be analyzed in learning situations (Mazeau and Pouhet 2014; Petitfour 2017).

Our fourth hypothesis is that the learning of figures must be based not only on the progressive construction of their geometric properties, mentioned above, but also on the pupil's analysis of their tracings, and the causes of their imperfections or possible errors.

9.3.2. *The SQUARE AND QUASI-SQUARE situation*

This situation is the only one that is proposed in both first and second grade, at the end of the second quarter. The material consists of four types of quadrilaterals: squares with sides of 4 cm; rhombuses with sides of 4 cm (with an offset of 2 mm); and two types of rectangles: "small" rectangles 4 cm by 3.7 cm and "large" rectangles 4 cm by 4.3 cm (Figure 9.6).

Figure 9.6. *Square, rhombus and rectangles*

Experiments have shown that these pieces are side by side too similar to be distinguished by first or second graders. The hypothesis is that, thanks to the assembly of these different shapes, the characteristics of the square and the specificities of the rhombus and the rectangle can be identified. The situation has three phases and associates the anticipation of actions on objects and the tracing of figures (squares):

– phase 1: identify the differences between squares and rhombuses;

– phase 2: identify the differences between squares and rectangles;

– phase 3: to implement the properties of the square by tracing it.

9.3.2.1. *Phase 1: distinction between square and rhombus*

The principle of this situation is to put in check findings based on perception (the pieces are close enough that their differing angles can, for many pupils, pass as inaccuracies that happened when cutting).

In the first problem, the pupils, in pairs, have to produce an assembly of four pieces, squares or rhombuses (Figure 9.7(a)) from a set of eight (four squares and four rhombuses) and then keep track by tracing the outline (Figure 9.7(b)). In a second problem, pupils are asked to construct a square using a combination of three squares and one rhombus. After several attempts, they find that it is impossible to produce one. Many pupils discover that the angles of these rhombuses are not the same, that they differ from those of the square, and that when a square is rotated, the assembly does not change, whereas with the rhombus a hole or overlap is formed.

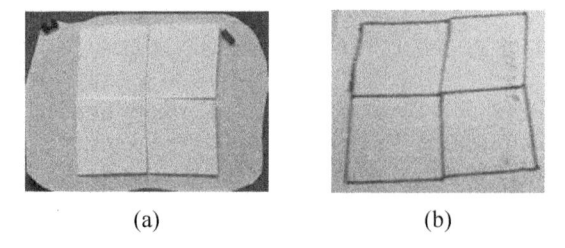

Figure 9.7. *(a) Assemblies and teaching aids; (b) an assembly layout*

9.3.2.2. *Phase 2: the distinction between a square and rectangle*

In this problem, pupils have two squares, a "small" rectangle and a "large" rectangle (in the first grade) and three squares, two "small" rectangles and two "large" rectangles (in the second grade). They must align these figures on the folded edge of a sheet of paper by wedging them, without leaving any space, so as to produce the longest possible assembly. During the group sharing stage, each pair come up to the front and show their proposal (using templates that are three times bigger), marking their solution with a line, and on their sheet (Figure 9.8).

Figure 9.8. *The assembly of squares and rectangles*

The pupils are thus led to express observations and put forward hypotheses on how to produce the longest (or shortest) strip possible, through the actions applied to the shapes: "When you turn it, it makes it bigger; it makes it longer"; or on their properties: "There is a side where it is long, and a side where it is short." Some pupils discover what differentiates a square from a rectangle: "If you turn them (the squares), it always looks the same"; "The square is not like the rectangle, it doesn't lengthen the line"; "They all have sides of the same size." This makes it clear that the sides of the square are of equal length and also, for some first graders, that the order of the pieces – identically oriented – does not change the length of the assembly.

9.3.2.3. *Phase 3: the construction of a square*

Pupils are asked to complete a square, of which only one side has been traced, using a non-graduated rectangular ruler whose 10 cm length is the same as the side of the square, thus relieving pupils from the task of measuring lengths. Many pupils, even in second grade, place approximately one side of the ruler along the line and not along the line itself (the angles are therefore not right angles). The sharing of information allows for the analysis of this error, by checking the right angles with the ruler, as well as distinguishes this error from simple clumsiness when tracing. The resumed drawing of the line allows for an improved positioning of the ruler, and precedes the institutionalization of the properties of the square that were formulated in the group sharing stage of the previous phases: a square with four "equal" sides (of the same length) and four right angles.

9.3.3. **The emergence of criteria for comparing solids: the IDENTIFYING A SOLID situation**

The hypotheses underlying this learning process are the same as those described above. The pupils are given two successive problems. In the first task, from a given collection of solid shapes (Figure 9.9), they have to find an object that has been shown to them by the teacher previously; in the second task, they need to determine an object that was hidden in a bag that they could only examine with their hands.

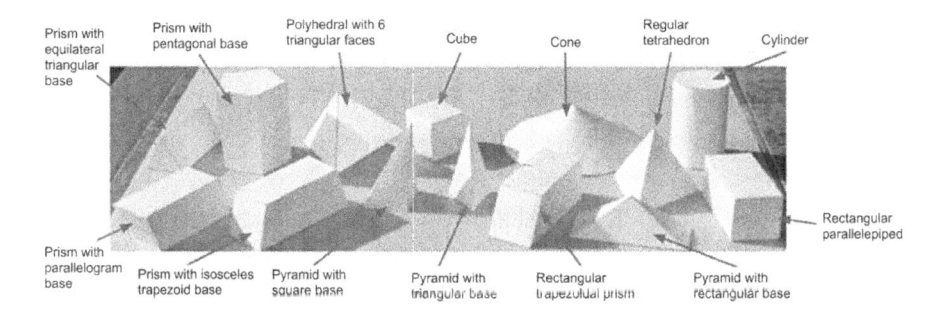

Figure 9.9. *The collection of solid shapes*

The solids are homogeneous in terms of their material (same material and same color). All are polyhedral (prisms and pyramids of various kinds) except two (the cone and the cylinder) which, to a certain extent, are interlopers. Polyhedra are divided into two families: prisms and pyramids.

Knowledge of specific terms is neither a prerequisite nor an objective. In both activities, pupils will form families, for example, by choosing prisms (Figure 9.10).

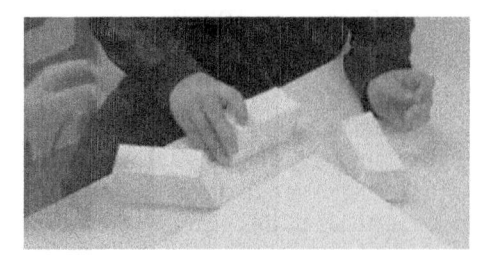

Figure 9.10. *Grouping solid shapes*

They will then go beyond questions of similarity, comparing, for example, "There it was tilted" and "There it was not tilted" (Figure 9.11).

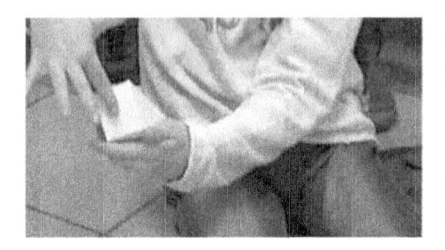

Figure 9.11. *Gestures associated with the formulations*

This situation allows for an evolution in the consideration of comparison criteria, which successively pass through (Coutat and Vendeira-Maréchal 2019):

– a global perception: the shape is expressed through an analogy with a familiar object;

– the formulation of a local spatial characteristic, for example: "There, it is pointed";

– a local analysis in geometric terms: "There are three sides here, and here there are four";

– a complete analysis of the geometric properties: counting the faces and vertices and sometimes the edges, characteristics of the faces, relations of among the faces, etc.

9.3.4. *Identification of cube properties: the CUBE AND QUASI-CUBE situation*

The problem posed is as follows: after examining a truncated cube missing one piece (Figure 9.12, top image), and consisting of seven small cubes, the task is to determine which of the four solids – consisting of one cube and three closely related solids – is an exact replacement for the missing cube. The pupils, in second grade, by comparing them successively two by two – and without having the large truncated cube in front of them – must determine which of the candidates can be used. Each pair must then justify, in writing, the reasons behind their choice: the relative position of the faces, the properties of the lengths and/or the edges, the angles of the faces, identifying faces of different sizes or shapes.

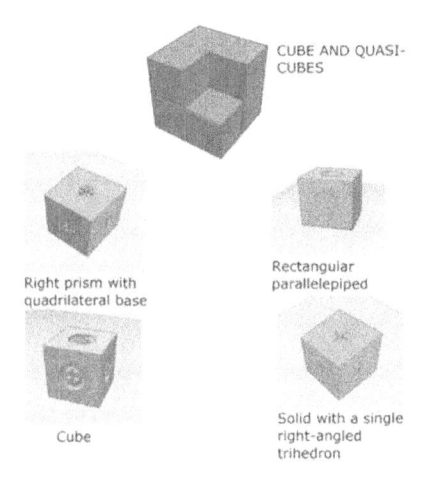

Figure 9.12. *Introducing quasi-cubes*

As we have seen for the problems of assembling figures, the gestures (turning around, turning over, superimposing, juxtaposing, nesting, etc.) reveal an intentionality: they have exploratory, control and evocation functions, etc. Also, rather than developing typologies that result from a predominantly visual observation of solids, we propose to make their properties emerge through the explicitation of their differences, such as they appear in the resolution of problems implying the assembling of objects in space or on the plane with similar shapes. This choice, which allows access to gestures and criteria for comparison produced by the pupils themselves, also concerns, as we have seen, the study of plane figures.

9.3.5. *Progression on solids and plane figures*

Our progression privileges the use of plane shapes and figures for solving problems involving solids, and reciprocally, the role of solids in stimulating the study of plane figures. This mutual contribution favors a change in the status of these objects, which are no longer simply objects of space or the plane, but are carriers of geometric relationships between the elements that compose them.

Figure 9.13 presents a chronology of all eight situations, four relating to objects in space (in bold) and four to those on the plane.

Situation	Level/Period	Problem	Institutionalized Notions	Solicited Notions
EMPRESSES	Grade 1 Period 1	Associate a solid with faces	Set of faces of a solid	Squares, rectangles, trapeziums
TO IDENTIFY A SOLID	Grade 1 Period 2	Characterize a solid by its faces	Vertex, face, dimension	Pyramids, prisms, etc.
TO IDENTIFY A SHAPE	Grade 1 Period 3	Characterize a shape by its dimensions	Straight dimensions Number of dimensions of a polygon	Triangles, squares, rectangles, trapeziums polygons, etc.
SQUARES AND QUASI-SQUARES	Grade 1 Period 5 Grade 2 Period 2	To distinguish the square from the close quadrilaterals To draw right angles	Properties of the square To draw right angles	Squares, rectangles, rhombus
CURVED FIGURES	Grade 2 Period 4	Analyze a figure	Circle as a figure with constant curvature	Precision of a trace
CUBE AND QUASI-CUBE	Grade 2 Period 4	Distinguish the cube from nearby solids	Properties of the cube	Cube, square, rectangle
WRAPPING A SOLID	Grade 1 Period 5	To associate faces to a solid	Wrapping a solid	Cube, straight slab, straight prism

Figure 9.13. *Activity progression*

The progression thus combines learning about solids and plane figures, as the problems proposed in the first two situations indicate.

9.4. The appropriation of research results by the resource[1]

The reliability of our teaching proposals also depends on the possibility for teachers to appropriate and implement them. It therefore assumes that the teacher's decisions are made explicit and that they have a progression justified by our experimentations. Our resource includes a detailed description of the situations and answers to questions about learning and teaching that teachers may ask themselves

---

1 The first 24 pages of the book, including the presentation of the organization for the resource, and that of the theme "Traces and use of straight line," as well as the situation "Stripes: and a description of its issues," are available at: hatier.fr/flip/flex/97822189988120?token=43e51085ff6aa79737712bc4507c88eb.

after their implementation. The learning is structured around three themes: "Objects in space and along the plane," "Tracing and using straight lines," which also initiates the analysis of figures, and "Locating." This resource also presents training activities, knowledge stabilization and additional insights into knowledge and learning processes or teaching choices.

This appropriation is supported by five features:

– the reliability of the situations, based on experimental results insofar as we are able to guarantee that the procedures described are those that the teachers will encounter;

– the analysis of the initial knowledge of the pupils, allowing the teachers to anticipate the procedures, the difficulties, the evolutions and thus to develop an estimation of pupils by analyzing their mathematical activity;

– the precise description of the teacher's interventions at different moments of each situation (formulation of the instructions, accompaniment in the research, management of the exchanges and conclusion drawn in the group sharing stage, etc.), thus providing indications on the professional gestures that ensure the smooth running of the situation, and allowing the teachers to anticipate their decisions (see Appendix 2);

– the provision of all materials (see Appendix 3) to try to limit the teacher's workload to a minimum in relation to the material preparation;

– the provision of tools for the teacher to analyze their practices, and the clarification of didactic issues in the form of questions for teachers and trainers, to make this tool a training tool.

The clarification of analytical tools, the formulation of hypotheses and their testing, are components of our research, and as such, can be found in these resources.

9.5. Conclusion

The acquisition of geometrical knowledge by 6- to 8-year-old pupils requires the analysis of their initial knowledge, resulting in particular from their familiarity with practices, which they resort to for the resolution of spatial or geometrical problems, in order to know, in particular, which procedures they engage with and which properties of the objects they use. Concerning the analysis of their productions, which is one of the investigations of this book, we have explained how the didactic situations proposed for the learning of straight lines, figures and solids, lead them to question erroneous conceptions, notably linked to perception, and which thus allow for an evolution in understanding (lines, constructions, gestures, linguistic formulations). These situations also bring about new criteria for judging the geometric knowledge of pupils.

In our opinion, the production of teaching resources presupposes that, in order for educators to appropriate a progression covering all the concepts in the program, their actions in carrying out situations should be described and they should have the tools to question the choices proposed to them as well as to analyze their own practices. Therefore, in our research, we have chosen to not separate the development of problems from the reflection of their implementation by teachers.

9.6. References

Berthelot, R. and Salin, M.H. (1992). L'enseignement de l'espace et de la géométrie dans la scolarité obligatoire. PhD Thesis, Université Sciences et Technologies Bordeaux I.

Coutat, S. and Vendeira-Maréchal, C. (2019). Reconnaissance de formes à l'école maternelle, un point de vue didactique et psychologique. In *Nouvelles perspectives en didactique : géométrie, évaluation des apprentissages mathématiques*, Coppé, S., Roditi, É., Celi, V., Chellougui, F., Tempier, F., Allard, C., Corriveau, C., Haspekian, M., Masselot, P., Rousse, S. et al. (eds). La Pensée Sauvage, Grenoble.

Douaire, J., Argaud, H.-C., Emprin, F., Frémin, M. (2020). *ERMEL géométrie CP-CE1*. Hatier, Paris.

ERMEL. (2006). *Apprentissages géométriques et résolution de problèmes en CE2-CM1-CM2*. Hatier, Paris.

Houdement, C. (2019). Le spatial et le géométrique, le yin et le yang de l'enseignement de la géométrie ? In *Nouvelles perspectives en didactique : géométrie, évaluation des apprentissages mathématiques*, Coppé, S., Roditi, É., Celi, V., Chellougui, F., Tempier, F., Allard, C., Corriveau, C., Haspekian, M., Masselot, P., Rousse, S. et al. (eds). La Pensée Sauvage, Grenoble.

Mazeau, M. and Pouhet, A. (2014). *Neuropsychologie et troubles des apprentissages chez l'enfant : du développement typique aux dys*. Elsevier Masson, Issy-les-Moulineaux.

Petitfour, E. (2017). Enseignement de la géométrie à la fin du cycle 3. Proposition d'un dispositif de travail en dyade. *Petit x*, 103, 5–31.

10

Tool for Analyzing the Teaching of Geometry in Textbooks

A parliamentary report on mathematics education (Villani and Torossian 2018) emphasizes the importance of library resources in the activities offered to pupils. Building on the study presented by Mounier and Priolet (2015), which highlights the privileged place of textbooks by primary school teachers, this report insists on the need for these textbooks to be "easy to use" while maintaining "an ambition for rigor and quality in [their] contents" (Villani and Torossian 2018, p. 56, *translated by author*). With this in mind, they conclude that there is a need to provide teachers with a tool that "allows them to make an informed choice, in light of a set of relevant criteria" (ibid., p. 57, *translated by author*). We therefore addressed this issue – providing tools to inform teacher choices – by focusing on the field of geometry. Middle school (9–12 years) is an important period in the development of geometric learning; we chose to focus on the fourth-grade level (9 and 10 years), the first year of this age bracket. We next focus our study on the teaching of the theme "perpendicularity and parallelism" to the fourth grade in France, during the time when these notions were introduced according to the French curricula.

Based on the definition of *textbook*[1], we employ the term to mean a given collection, the various documents intended for the pupil (pupil book, file, exercise book, repertory, etc.) as well as for the teacher (teacher's guide or supplementary information transmitted by the editor via their website or in further works).

Chapter written by Claire GUILLE-BIEL WINDER and Edith PETITFOUR.
1 Decret 2004-922 of August 31, 2004. Avalaible at: https://www.legifrance.gouv.fr/jorf/id/JORFTEXT000000445213.

The work of Remillard (2010) emphasizes the importance of the form and appearance of this "curricular material" and the way in which teachers make use of it. Therefore, building on a previous work presenting a didactic analysis of digital textbooks (Guille-Biel Winder and Petitfour 2021), we took this aspect into account by studying, among other things, the medium (the form in which the resource is disseminated), as well as the voice, which corresponds to "the way in which the authors' discourse is represented, and the way in which they communicate with the teacher" (Remillard 2010, p. 106, *translated by author*). In this study, we focus on geometric knowledge, seeking more specifically to determine what might inform, from a didactic point of view, the choice of textbooks as teaching aids.

The first part introduces the general framework and the theoretical tools which we rely upon. The second part is devoted to the presentation of the criteria, and the third part details the analysis grid for a textbook.

10.1. General framework and theoretical tools

We place ourselves in the framework of the Anthropological Theory of Didactics (Chevallard 1999). We take as a basis the didactic co-determination scale (Chevallard 2002), which we explain, after which we present the mathematical organization of reference (Bosch and Gascòn 2005), as well as the theoretical tools which allow for the analysis at the level of a specific mathematical organization (Chevallard 1999).

10.1.1. *Didactic co-determination scale, mathematical and didactic organizations*

The didactic co-determination scale (Chevallard 2002) highlights the interdependence of its different levels in terms of conditions: "Each level of this scale is the locus of origin for certain conditions that often appear as constraints at other levels." (Chevallard 2011, p. 12, *translated by author*). Indeed, the teaching of mathematical knowledge, as proposed by a textbook, is subject to these multiple constraints.

We focus our textbook analyses at different levels of the co-determination scale: the discipline (mathematics), the field of study (geometry), the area of study (recognizing and using certain geometric relations), the topic of study (perpendicularity and parallelism relations), the topics of study (identifying perpendicularity or parallelism relations, making tracings of lines and segments corresponding to perpendicularity or parallelism relations). We also consider less-specific levels of mathematics: the school level and the pedagogical level. The first one corresponds to "the level of constraints and points of support that have to do with the school institution itself" (Chevallard 2002, p. 13, *translated by author*). We focus at this level on the origin of constraints linked to official instructions, as well as those linked to the material organization (e.g. multi-level courses) or the temporal organization of the year (e.g. alternating work periods due to school vacations). The second includes issues and constraints that affect school studies. In particular, the temporal constraint for the length of a session has an impact on the time devoted to the study of a given subject. Pedagogical choices are also made at this level. Some of these may have a theoretical background cited in the textbook, such as learning theories (e.g. constructivism (Piaget 1964) or pedagogical trends: explicit pedagogy (Rosenshine 1986), spiral pedagogy (Bruner 1960), etc.). Figure 10.1 shows the schematization of Chevallard's (2002) didactic co-determination scale, contextualizing it for our study.

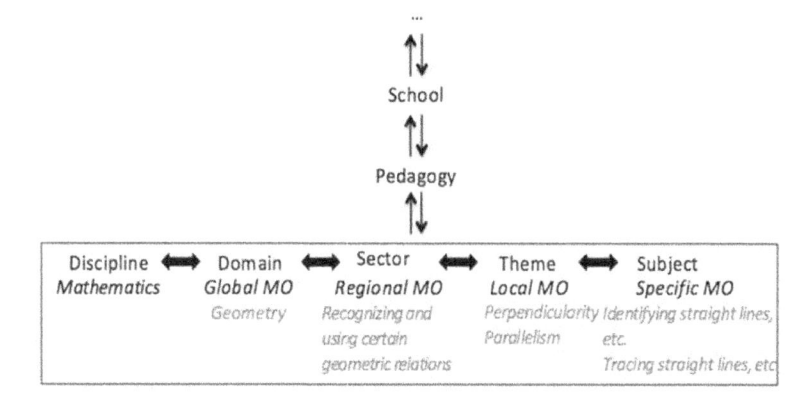

```
                          ...
                          ↕
                        School
                          ↕
                       Pedagogy
                          ↕
┌─────────────────────────────────────────────────────────────────────────┐
│ Discipline  ⇔  Domain   ⇔   Sector        ⇔  Theme       ⇔  Subject    │
│ Mathematics    Global MO    Regional MO      Local MO       Specific MO │
│                Geometry     Recognizing and  Perpendicularity Identifying straight lines,│
│                             using certain    Parallelism    etc.        │
│                             geometric relations              Tracing straight lines, etc│
└─────────────────────────────────────────────────────────────────────────┘
```

Figure 10.1. *Extract from the didactic co-determination scale*

Moreover, we take into account different scales concerning the teaching of mathematical knowledge: specific, local, regional and global mathematical organizations (MO) (Chevallard 1999).

Task type and technique are the two components for the know-how of specific MO. A type of task is defined as any type of activity thought of as elementary, in the sense that it could be stated using an action verb and an object complement. A technique is a precise way of doing things that allows tasks of a certain type to be carried out. The other components of the specific MO (technology and theory) relate to mathematical knowledge. Now, this knowledge is evoked through the manipulation of ostensives (Bosch and Chevallard 1999), that is, objects that have a material form: tangible objects (material ostensives), gestures (gestural ostensives), words and discourse (linguistic ostensives), diagrams, drawings, graphics (graphical ostensives), writing and formalisms (scriptural ostensives).

The highlighting of mathematical and didactic organizations as well as the influence they have on each other, leads us, on the one hand, to identify the points on which we focus the analysis of the proposal for teaching geometric knowledge in a textbook – in relation to a reference MO that we will determine – and, on the other hand, to take into account different levels of the didactic co-determination scale in the analysis of each of these points. We thus focus the analyses on the types of tasks proposed, the techniques called upon, the knowledge in play, the ostensives manipulated and, finally, on the organizational and planning elements (Guille-Biel Winder and Petitfour 2018), that is, the elements of the textbook that explain and/or testify to the organization and planning that has been chosen. Therefore, these analyses focus on a specific level of the didactic co-determination scale or rely on a back and forth between different levels.

10.1.2. Reference MO and theoretical tools for analysis

In this section, we present the reference MO constructed from an analysis of the current middle school programs and its different declinations in the textbooks for the level being studied. We also explain the theoretical tools used for the analysis of the subject of study. The official texts taken into account are those available to teachers on the Eduscol website[2].

These texts include the school timetable, the curriculum for the consolidation cycle (MEN 2015) as well as the adjustments made for the start of the 2018 school year (MEN 2018a), the accompanying resource on "Space and Geometry in Middle School" (MEN 2018b), and finally the end-of-year expectations and yearly progression benchmarks (MEN 2019).

10.1.2.1. Types of tasks and techniques

We distinguish between two main categories of elementary tasks as they relate to instrumented constructions involving the perpendicularity and parallelism relations: those associated with the recognition of the relation, and those related to the construction of lines verifying this relation. In the first category, we dissociate the types of tasks in which the pairs of lines are given (verification of the relation), from those in which they are not (identification of the relation). Concerning the second category, we identify three variants, depending on the constraints of the line(s) to be traced: no line is given, a line is given, a line and a point are given. Moreover, for the perpendicularity relation, when both a line and a point are given, we take into account the following didactic variable: the given point belongs to the line given, or not. For the parallelism relation, we take into account the following didactic

---

2 https://www.education.gouv.fr/programmes-et-horaires-l-ecole-elementaire-9011.

variable: the distance between the two lines is given, or not. Table 10.1 provides a summary of the types of tasks.

Basic task types related to instrumented constructions	
Recognition of the perpendicularity/parallelism relation	Check the relation between two given lines.
	Identify the relation between two lines in a network of lines.
Construction of line(s) verifying the perpendicularity/parallelism relation	Construct two lines that verify the relation. *Variable for the parallelism relation*: the distance between the two lines is given, or not.
	Construct a line verifying the relation with a given line. *Variable for the parallelism relation*: the distance between the two lines is given, or not.
	Construct the line verifying the relation with a given line and a given point. *Variable for the perpendicularity relation*: the given point belongs to the given line, or not.

Table 10.1. *Types of elementary tasks involving the relations of perpendicularity and parallelism*

This study allows us to identify the types of tasks expected in the official texts. Types of recognition and construction tasks corresponding to each of the two relations are mentioned in the 2015 French programs.

This was the same with the 2018 adjustments, however, with the additional precision for the tracings, to be perpendicular/parallel to a given line while passing through a given point. Furthermore, perpendicularity is mentioned as expected at the end of the fourth grade, the possibility that the given point is external to the line, which suggests taking into account the didactic variable, "position of the point in relation to the given line", with the values "point on the line" and "point outside the line" in the types of tracing tasks being proposed. For parallelism, the variable, "distance between the two lines", is not mentioned. The resolution of the types of relation recognition or tracing tasks requires different techniques for which the official instructions are not very precise. The 2015 programs give "examples of instruments": "graduated ruler, set square, compass, angle template, tracing paper, strips of paper"; "medium": "geoplan, grid paper, dotted paper, plain paper"; and "materials": "pencil and paper, dynamic geometry software, introduction to programming, software for viewing maps, and/or plans". In the resource document linked to the 2018 adjustments, the task type "recognize" is defined as "identify, perceptually, using instruments, or using definitions and properties" (MEN 2018b, p. 5). We can therefore consider recognizing relations by visual perception, by "reading" on a squared or pointed medium, by using a set square, a set square and a graduated ruler, or even other artifacts not cited by the programs but that can be part of the classroom resources (such as the rod guide or the rule-square[3]). The instrumental techniques to be developed are, in particular, those that make use of a set square and a ruler.

10.1.2.2. *The knowledge at stake in instrumented actions*

According to the framework for analyzing instrumented action (Petitfour 2017), different levels of knowledge are involved in solving the types of geometric recognition and tracing relation tasks previously mentioned, namely geometric, spatial, graphical, technical and practical knowledge.

Geometric knowledge concerns geometric objects, properties and relations. It can be identified through vocabulary and language formulations. This knowledge includes, in particular, the meanings underlying the instrumental techniques implemented in the instrumental actions. For our analyses, we rely on the following meanings for perpendicularity and parallelism relations, which are accessible in middle school (Reymonet 2004; Dussuc et al. 2006; ERMEL 2006).

Two perpendicular lines can be viewed primarily as two straight lines that intersect forming four right angles/two straight lines that intersect forming a right angle, related to the notion of an angle; a straight line of a given direction, "leaning neither to one side nor the other" with respect to the other; the straight line passing through a given point, which makes it possible to obtain the distance from this point to a given straight line, related to the notion of distance; straight lines obtained by folding a sheet of paper in half, in connection with the notion of symmetry; straight lines supporting consecutive sides of a rectangle in connection with an understanding of rectangular properties; straight lines that have a relation between their slope (their product is equal to -1) that is unexpected, but which can be observed on a grid support when the straight lines pass through the nodes of the grid (Figure 10.2).

Two parallel lines can be seen primarily as in Figure 10.3: non-intersecting (or never intersecting) lines in connection with incidence; lines of constant distance to be related to the notion of distance; lines of the same direction or lines of the same slope, in connection with the notion of angle; lines obtained by translation, with reference to transformations of the plane; lines perpendicular to a third line, explicitly based on the notion of perpendicularity, and which can be considered as a special case of same direction lines; lines supporting opposite sides of particular quadrilaterals (the square, the rectangle, even the trapezoid or parallelogram).

---

3 A rule-square is both a ruler and a set square.

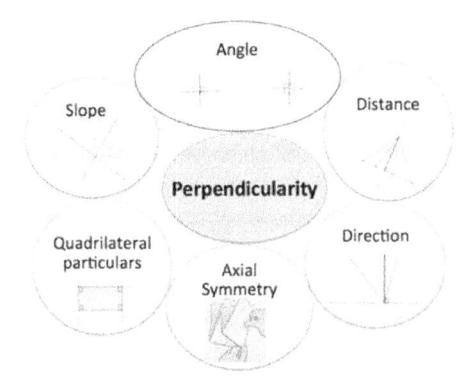

Figure 10.2. *Presentation of the different meanings intrinsic to perpendicularity relations*

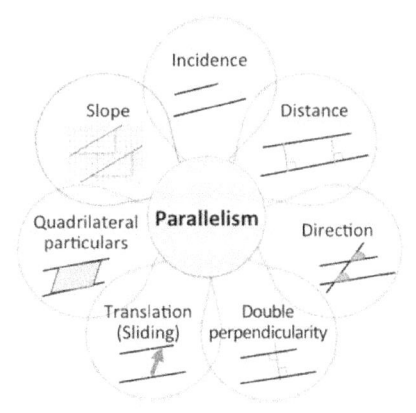

Figure 10.3. *Presentation of the different meanings intrinsic to the parallelism relation*

The official instructions provide little information on the meanings of the relations to be addressed at the beginning of the fourth grade. In the 2015 French programs, the signification for the notion of perpendicularity linked to the notion of an angle can, however, be inferred from the use of the set square, mentioned for the tracing task: "Trace with the set square the perpendicular line so that a given line passes through a given point." The one related to distance is explicitly mentioned, and the notions of parallelism and perpendicularity are linked to one another: "determine the shortest path between a point and a line (…), in connection with perpendicularity". Therefore, for the relation of parallelism, the signification "constant distance" appears to be advanced. However, the wording of the programs "construction of parallel lines, link with the property linking parallel and perpendicular lines" can also refer to the technique of sliding the square along the ruler, associated with the terms "double perpendicularity" or "translation". In the 2018 adjustments, the notions of parallelism and perpendicularity remain implicitly linked by the imposed use of the set square in the tracing of parallel lines, making the previous meanings for parallelism possible again: "constant deviation", "double perpendicularity" or even "translation". The first one seems again to be the most likely in the hypothesis for a coherence of programs between the construction technique approached and the underlying signification being studied, since the notion for the distance of a point to a line still appears in the programs, and double perpendicularity is only expected at the end of sixth grade (11 and 12 years old) (MEN 2019). Finally, the signification for parallelism and that of perpendicularity in relation to particular quadrilaterals appears to have been addressed as well, since the relations have to be "employed".

Spatial knowledge is related to the experience that the subject has of the tangible space. It is related to the perceptive selection of spatial information and to its interpretation, to the anticipation of transformations and displacements. Rouche (2008) emphasizes that the notions of right-angle, perpendicularity and parallelism are initially intuitive and practical, but not scientific. It is familiar to human beings in the common sense, thanks to the privileged physical directions of vertical and horizontal: the vertical lines exist in the everyday environment due to the force of gravity and have a concrete existence (poles, chandelier suspension cable, tower, etc.) and the horizontal ones exist as straight lines on horizontal planes (straight road in the desert, edge of a table, etc.) The spatial knowledge that results from this apprehension of the environment leads to a perception of the right angle as an angle formed by a horizontal side and a vertical side (consecutive edges of a window, a painting, a tile, etc.), or as a "corner" of a given rectangular shape (corner of a table, a wall, etc.), and – to a lesser extent – a perception of parallel lines as vertical or horizontal lines. It should be noted that spatial understanding could be considered as a support for the construction of a perceptual invariant of spatial positions for two lines, but that it can also present itself as an obstacle in the learning of geometric relations (ERMEL 2006). The latter will indeed be less easily recognized in a non-prototypical orientation (oblique lines). Moreover, recognition will be less easy if the lines to be considered are inaccessible in a network of lines or on a complex figure.

Graphical knowledge leads to the discernment of relevant graphical information, extracted visually from the drawing, and allows for the interpretation of its geometric meaning. It is related to graphical features (tracing and coding) and associated scriptural features, such as notations (e.g. the capital letter written next to a point to name it), as well as symbols (e.g. $d // d'$ signifies that the lines d and d' are parallel). Another example: "a line is represented as a straight line that can be extended as far as you want" is a graphical knowledge to be mobilized, especially for lines that represent two perpendicular lines (Figure 10.4(a)) or non-parallel lines (Figure 10.4(b)) do not intersect on the graph. Assimilating the lines to their representation does not lead to the identification that the lines are in fact intersecting.

Figure 10.4. *(a) Perpendicular lines and (b) non-parallel lines*

Technical knowledge concerns the functions and patterns for the use of instruments (Rabardel 1995). It is linked, for a given instrumented action, to the relation between the parts of the instrument and the graphical objects representing the geometric objects being considered. For example, the function of the set square is to verify or produce a right angle. To draw a half-line perpendicular to a given line d, originating at a point A on the line d, one side of the right angle of the set square is placed on the line d with the vertex of the right angle of the set square placed on the point A, then the second side of the right angle of the square is traced from the point A.

Practical knowledge is related to the material and physical plan linked to the instrumented action in connection with manipulative skills (the coordination of movements and postural adjustments needed to manipulate the instrument with dexterity and precision). It also concerns its organization (the ability to plan a sequence of actions according to a set plan).

10.1.2.3. *Ostensives*

Our analysis focuses on different ostensive objects (Bosch and Chevallard 1999): those in the register of materiality, such as tangible objects, those in the register of trace such as graphical (tracings, encodings), scriptural (notations, symbols) and linguistic (vocabulary and language expressions) ostensives.

We pay particular attention to the objects used to teach geometric relations (Guille-Biel Winder and Petitfour 2019): these allow us to study the possible link with reality as well as the ways in which the transition into geometry is being implemented. Based on the study of about 20 fourth-grade textbooks on the market during the 2017–2018 school year, we have highlighted a variety of objects for which perpendicularity and parallelism relations are worked upon.

We have organized these into two broad categories (Figure 10.5): objects of the material world[4] (Laparra and Margolinas 2016) or their figurative representations, by uses known to pupils, belonging to the pupils' everyday environment; graphical objects correspond to the material objects (drawings) of school geometry (Houdement 2007). In the first category, we distinguish objects from around the world that are culturally representative of spatial relations from those that are not. Examples of world objects that are culturally representative[5] for the relation of parallelism are tire tracks, referring here to the relation of incidence; railroad tracks that follow straight lines, accompanied by their crossings that further reference the property of constant deviation; the edges of a flat ruler referring to opposite sides of a rectangle. In the second category, we distinguish between representations of geometric objects and models of worldly objects. Objects that are culturally representative of the spatial relation are sometimes modeled in the graphical space under the same representations as geometric objects. These representations can then play the role of pivot signs (Bartolini Bussi and Mariotti 2008) since they can allow for a transition towards the geometric.

Objects of the Material World		Graphical Objects	
culturally representative of spatial relations	non-specific representative of spatial relations	modeling a world object	representative of a geometric object

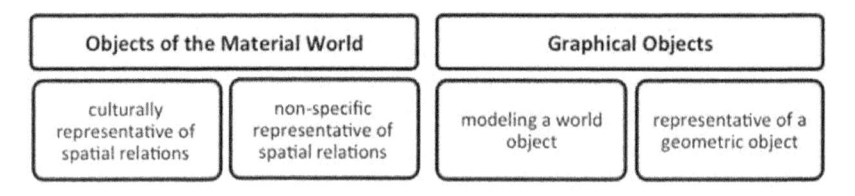

Figure 10.5. *Four categories of objects through which to teach geometric relations*

Concerning symbolism, we identify in official instructions the notations to be approached in fourth grade: "the notations representing parallelism (//) or perpendicularity (⊥) are introduced as they are used and not at the beginning of a learning process" (MEN 2018b, p. 11, *translated by author*). These notations are therefore not to be introduced in fourth grade, at the time when the notions of perpendicularity/parallelism are being introduced. Finally, only the notation "segment [AB]" is encountered in fourth grade without being required.

With respect to language ostensives, we use tools for the logical analysis of mathematical concepts (Petitfour and Barrier 2019). Perpendicularity and

---

4 "An object that is most often material, present both in and out of school, and that as a result evokes for the pupils the affects and uses that he or she already knows" (Laparra and Margolinas 2016, p. 169, *translated by author*).

5 Representative (adjective): that which is used for representation. The hieroglyphs were signs representing visible objects (Littré 1876). Website: http://www.la-definition.fr/definition/representateur, visited 20/12/2018 (*translated by author*).

parallelism are geometric relations, while the concept of a right angle is a geometric property for an angular sector. Logically, perpendicularity and parallelism can then be analyzed as binary relations between two lines: we say that "two lines d and d' are perpendicular/parallel" or that "line d is perpendicular/parallel to line d'" or that "line d' is perpendicular/parallel to line d" (symmetry of the relation). A line d being given, there is an infinity of lines that are perpendicular/parallel to it. This property is expressed in language by an indefinite article when a line d' is introduced: "The line d' is a line perpendicular/parallel to the line d". The relation of perpendicularity/parallelism associated with a relation of belonging links two lines and a point. In this case, we have a conjunction of two binary relations of the type "d is perpendicular/parallel to d' and M belongs to d'", formulated with the following syntax: "d' is the perpendicular/parallel line to the line d passing through (and which passes through) the point M", the definite article "the" expressing both the existence and the uniqueness of the line d'.

Having established the general framework and the theoretical tools for our analyses, we introduce in the following section the criteria that guide the analysis.

10.2. Analysis criteria: definition and methodology

In relation to the constraints to which the teaching of mathematical knowledge proposed by a textbook is subject to, we have broken down our analyses according to five criteria: institutional conformity, educational adequacy, relevance of teaching the proposed knowledge, coherence with respect to the knowledge taught and validity of the proposals in relation to mathematics. The last three criteria reflect the didactic quality of the textbook.

10.2.1. *Institutional conformity*

Textbooks are situated in the school teaching context of the country in which they are produced and used. Consequently, it is necessary to question the conformity of textbook proposals to the official instructions transmitted by the institution concerning the knowledge and skills to be taught, as well as concerning the organization of the classroom (in terms of the co-determination level constraints of the school in question).

The analysis for the institutional conformity of a textbook is based on the OM of reference, achieved by comparing the tasks, techniques, significations of the expected relations, the desired ostensives and the organization advocated within the textbook's proposals.

10.2.2. *Educational adequacy*

The design of a textbook is also driven by the pedagogical choices of its author(s). It therefore seems important to examine the adequacy of the theoretical support claimed, and the proposed course of action. The analysis of the educational adequacy takes place in two steps. The first consists of identifying, with the various resources available (the teacher's guide, the introduction to the pupil textbook, the companion website, a book, etc.), the central ideas on which the author(s) claim to rely. The second is at the local level: we seek to determine whether these central ideas are put into practice within the curriculum (programming, teaching of knowledge – tasks, techniques, introduction of knowledge – the position of the ostensives), and even within the pedagogical and didactic recommendations that accompany it.

We note that educational adequacy is sometimes difficult to analyze objectively, since it depends on the authors' interpretation of current pedagogy. Sometimes, these supporting citations are not always made explicit by the authors of the textbooks.

10.2.3. *Didactic quality*

The teaching proposals are related to mathematical organizations (even if the other levels interfere). The three criteria that give an account of the didactic quality of the textbook are therefore based on a local analysis and a perspective with reference to MO.

Relevance to the teaching of knowledge concerns the didactic choices made by the author(s) concerning the chosen course (back and forth from the specific level to the global level of the MOs), the mathematical tasks and the choice of objects which they focus upon, the significations addressed, the presentation of techniques and language formulations (local and specific levels of MO). The study on relevance requires the identification of these choices, followed by their analysis from a didactic point of view.

The mathematical validity of the knowledge statements concerns the mathematical knowledge proposed to the pupils by the textbook (in connection with the constraints of the discipline). It is interrogated by studying the language formulations and symbolic notations used, as well as by analyzing the domain for the validity of concepts, and by the use of the proposed artifacts (local and specific MOs).

The coherence of the textbook in relation to the knowledge taught is addressed on two levels. At the level of local MOs, the links between the type(s) of task(s) proposed in the first encounter with the notion (perpendicularity and parallelism relations are introduced in fourth grade), the meaning(s) involved, the technique(s) given and the tasks proposed in the different activities of the textbook are examined. At the global and regional MO levels, the organization of knowledge adopted to work on the concepts is assessed. Coherence exists when the sequence is (chrono)logical, that is, when the concepts and/or techniques used have been learned beforehand (in the textbook or from the collection of previous years). Analysis at the global level hence requires the organization of knowledge to be explicated.

The analysis of a textbook according to the five criteria outlined above is operationalized by focusing on the types of tasks proposed and the techniques and associated ostensives, on the knowledge involved, and finally on its organization (through organizational and planning elements), while moving back and forth between the different levels of the co-determination scale. The following section presents this operationalization.

10.3. Introducing the analysis grid

The analysis grid corresponds to a breakdown of the analysis criteria for each of the five points of attention identified.

10.3.1. *Analysis of tasks and task types*

The uncovering of elementary task types (reference MO) allowed for the identification of the types of tasks to be taught. We are then in a position to question the criterion for the institutional conformity of the textbook. In order to meet this criterion, a textbook must propose resolutions for the following *types of tasks*:

– recognize the relation of perpendicularity;

– trace a perpendicular line along a given line passing through a given point belonging to the line/external to the line;

– recognize the relation of parallelism;

– trace the parallel line to a given line through a given point.

The proposed mathematical tasks meet the criterion for educational adequacy when they correspond to the authors' stated intentions. This is the case, for example, when the authors claim to rely on problem solving to conduct learning, and when the relations are introduced by consistent problems organized in such a way as to allow the pupil, beyond simple manipulation, to try things out, to find, to put forward hypotheses, to test them, to argue. This is also the case when the link with reality is provided by a textbook that advocates the lived/represented approach.

The relevance of the tasks is studied according to the reference MO, with regard to the choice of variables, such as, for example: the orientation of the lines or the complexity of the figure (bringing into play spatial knowledge), the need, or not, to extend them (bringing into play graphical knowledge), the instruments or the support – plain or squared – available (bringing into play technical knowledge). The mathematical tasks are relevant to the teaching of knowledge if this choice leads to the overcoming of obstacles and the understanding of concepts. A contrario, if the choice of variables does not allow for the overcoming of obstacles (e.g. when a qualitative property, for example, the same color, is sufficient to identify the parallelism relation), or the comprehension of concepts (e.g. when the parallel lines are in a privileged direction), the relevance criterion is not verified.

The criterion for coherence with the knowledge taught is verified when the tasks proposed in the various activities of the manual (practice exercises, "to go further", etc.) require the use of the techniques presented and/or institutionalized. When the tasks proposed call for techniques not covered, the criterion is not verified.

10.3.2. *Analysis of techniques*

We identify the techniques for recognizing or tracing relations that are employed by the textbook (institutionalized, explicitly addressed or even underlying), and next, we use the reference MO to assess the criteria for institutional conformity, educational adequacy, as well as relevance and coherence with respect to the knowledge being taught.

When, among the instrumented techniques proposed by the textbook, we find those that use a set square and a ruler, the textbook meets the criterion for institutional conformity.

When they are linked to the declared pedagogical approach, then the criterion for educational adequacy is verified. This is the case, for example, when the authors claim to rely on problem solving, and propose phases for the manipulation of instruments that allow for the development of the problem solving techniques being proposed.

The (first) presentation of the techniques is relevant, in relation to the teaching of knowledge, when a set of information necessary for the realization of the technique is provided, promoting: technical, spatial, graphical and practical knowledge. A contrario, the criterion is not verified when implicit information and/or useful spatial, graphical and practical knowledge is not present.

When the justification of the technique is based on one of the significations of the relation introduced, then the criterion for coherence in relation to the knowledge being taught is verified. This is not the case when the technique is based on an unexplained signification.

10.3.3. *Analysis of knowledge*

To carry out the analysis, we rely on significations for the relations of perpendicularity and parallelism that are accessible at the school level in question, identified in the reference MO (see section 10.1.2). After having explained the significations at stake in the textbook (those which are institutionalized, explicitly addressed or even underlying), we compare these to the expected significations (the criterion for institutional conformity). Finally, we analyze them according to the criteria for educational adequacy, mathematical validity, relevance and coherence with respect to the knowledge being taught.

The textbook meets the criterion for institutional conformity when the signification of the perpendicularity relation, linked to the distance between a point and a line, and at least one signification of parallelism in connection with perpendicularity are addressed. A contrario, if the signification related to distance for perpendicularity and/or the signification related to perpendicularity for parallelism is not addressed, then the textbook does not meet the criterion.

When the implementation of the notion, as proposed in the textbook, is in line with the learning theory or pedagogy to which the authors claim to refer (explicit, constructivist, behaviorist, etc.), then the introduction of knowledge is consistent with the authors' stated intentions. For example, when the authors claim to rely on problem solving in learning, and the first encounter proposed by the textbook with each of the relations corresponds to solving problems that give meaning to these notions, then the criterion for educational adequacy is verified.

The choice of meanings addressed is relevant to the teaching of knowledge if it leads a priori to an initial understanding of the relation. A contrario, the use of very varied significations in the time allotted for the teaching of the relation, does not allow for a good understanding of the relation: as such, it is therefore not relevant.

When the significations of the relations remain within their domain of validity, the criterion for mathematical validity is verified; otherwise, it is not. For example, the transposition from the plane to space regarding the signification of parallelism "lines that never intersect" is outside the domain of validity for this particular signification.

We note a coherence between the first encounter with the notion proposed by the textbook (in the form of "preparatory activities", "introductory situations") and the resulting institutionalization (emphasized by headings such as "let's remember", "memo" and/or proposed in the teacher's guide as a verbal formulation, in particular), if at the end of the first encounter, only the significations and/or techniques that have just been approached are institutionalized. The coherence criterion is not verified when at least one signification or technique is institutionalized without having been the subject of a first encounter.

10.3.4. *Analysis of ostensives*

Our analysis focuses first on notation and mathematical symbols (scriptural ostensives) according to the criteria for institutional conformity and mathematical validity. We identify the objects used to teach geometric relations (material ostensives) as well as the objects on which the work is focused (material or graphical ostensives). We analyze these with respect to the criteria for adequacy and relevance. The linguistic ostensives (words and discourse) are examined in light of the criteria for adequacy, relevance and validity. Finally, we analyze the instruments with regard to the criterion for validity.

The textbook meets the criterion for institutional conformity as long as any symbolism is not the object of learning.

When the proposed ostensives (language, instruments, objects) demonstrate the implementation of the declared approach, the textbook is consistent with the declared intentions of the authors and the criterion for educational adequacy is verified.

The choice of objects upon which the tasks are based is relevant to the teaching of knowledge if it leads to a valid representation of the relation and/or the geometric objects involved, or even if it promotes the understanding of the relation and/or one of its properties. In the activity shown in Figure 10.6(a), the reference to the Earth's parallels (imaginary concentric circles organized in a network around a sphere) leads to an erroneous representation of the notion of parallel lines. In the activity shown in Figure 10.6(b), the choice of objects (straws) is not very relevant because it can lead to problems of understanding if we interpret the arrangements of straws in a tangible manner (it looks more like the straws in examples 2–6 are touching each other, while in examples 1 and 3, the straw has been placed on top of the others, in which case they do not "intersect").

Yuna and Kim observe the trajectory of a satellite on a planisphere. They wonder how many parallels the satellite will have cut when it is gone around the world.

a) Can you help them?

b) Look for the name of the lines perpendicular to the parallels on a planisphere.

The "parallels" on a planisphere are all the straight lines parallel to the equator.

▷ Here are the straws that Lisa's group received.

a. **Write** straws that cut into a right angle.

b. **Copy** the table and **classify** the straws.

Perpendiculars	Non-perpendicular
...	...

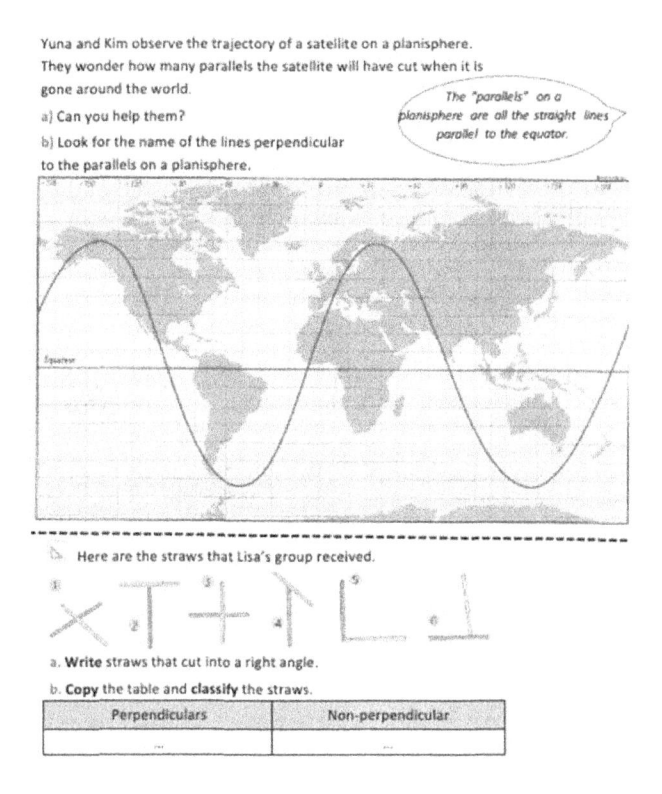

Figure 10.6. *Activities from two textbooks: (top) La Tribu des maths CM1 (Duplay and al 2008, p. 36) (bottom) J'aime les maths (Bourreau and al 2016, p. 144) (translated by author)*

Written Trace

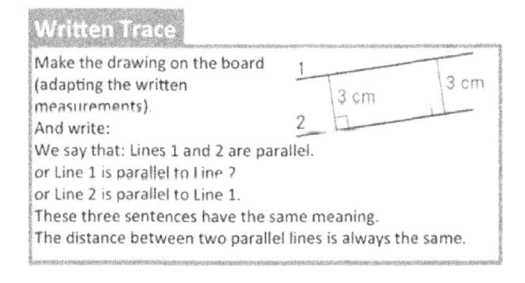

Make the drawing on the board (adapting the written measurements).
And write:
We say that: Lines 1 and 2 are parallel.
or Line 1 is parallel to Line 2.
or Line 2 is parallel to Line 1.
These three sentences have the same meaning.
The distance between two parallel lines is always the same.

Figure 10.7. *Teacher's guide of Nouveau Cap Maths Fourth-Grade (Charnay and al 2020b, p.194) (translated by author)*

The language formulations on the relation (of parallelism/perpendicularity) are relevant when the expression of the relation is decontextualized and if the equivalence between the formulations expressing its symmetry is institutionalized. This is the case, for example, in the written trace presented in Figure 10.7.

The use of an instrument is mathematically valid if the instrument is appropriate for graphically producing the desired geometric property, and thus obtaining a correct figure (Petitfour 2017). For example, the use of a perceptually positioned ruler to draw the second side of a right angle, from one side, is not valid, while the use of a correctly positioned set square to make this trace is.

The notations and symbols used are mathematically valid when they are conventional and correct with respect to usage. Therefore, in Figure 10.8, we consider the coding of the parallelism of two lines with arrows (even if it is in use in English-speaking countries) mathematically invalid.

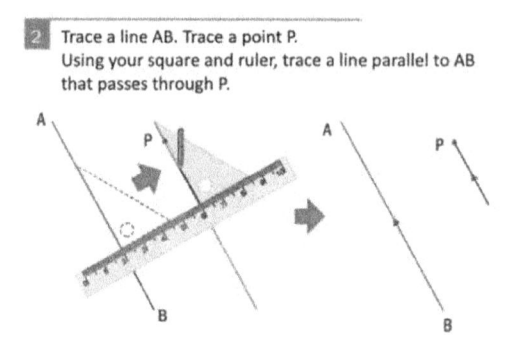

Figure 10.8. *The fourth-grade Singapore Method (Tek Hong 2009, p.83) (translated by author)*

Language formulations relating to perpendicularity and parallelism relations are mathematically valid when geometric language is used appropriately. For example:

– The uniqueness of the relation is expressed with the use of a definite article; on the other hand, the use of the article "a/an" (instead of "the"), for example, in the formulation "a line perpendicular to the line d passing through point M", invalidates the formulation.

– The terms "perpendicular" and "parallel" are used to express a relation between two lines; the presence of these terms referring to a property of a line ("the parallel line") therefore invalidates the formulation.

– The relations are indeed expressed in geometric terms. On the other hand, the mixing of everyday language and geometric language to express the relations – such as the expression "straws that intersect at a right angle" (see Figure 10.6(b)) – invalidates the formulation.

– The presentation of geometric concepts is mathematically correct. Using, for example, the expression "right angle" to refer to the "vertex of the right angle" reveals a lack of understanding of the concepts and invalidates the formulation.

10.3.5. *Analysis of organizational and planning elements*

In their proposals, the authors of primary school textbooks take particular account of the time constraint in organizing the teaching of mathematical concepts in the *curriculum*. This organization appears in terms of the temporal structuring of the knowledge covered at different levels (the division of the year into units of sequences, sessions, phases) and in relation to learning objectives. The organizational and planning elements associated with mathematical organizations (MOs) are thus expressed in terms of annual programming of the domain studied (Global MO) and of the progression relative to the teaching of the theme considered within this domain (Local MO). At the level of the discipline, we identify the distribution of mathematical content in the school year, and study the place reserved for the teaching of the field of study (Geometry) relative to other mathematical knowledges in the textbook (based on sequence plans, scenarios to be used). In a back and forth between the levels of the domain, sectors, themes and subjects of study, we analyze the articulation of the notions for perpendicularity and parallelism within the geometry sessions, with regard to the implementation of the tool/object dialectic (Douady 1986) relative to the types of tasks being proposed.

At the end of this work, we produce a synoptic table in order to generate an overview for the distribution of geometry teaching in the year, and on the consideration of the tool or object aspect of the concepts taught for the perpendicularity and parallelism relations. This table is based on the organization of school time in primary school[6]. We represent the distribution of the 36 weeks, according to the five work periods of comparable length (separated by the school vacations), on a rectangular strip. We place rectangles with areas proportional to the duration of the sessions devoted to the areas of studying geometry (plane figures: gray hatching; solids: grey waves; symmetry: gray area) and to the theme of study (relations of perpendicularity – dots on a background – and parallelism – black stripes). The object aspect of the relations is highlighted by white dots on a black background for the relation of perpendicularity and vertical stripes for the relation of parallelism, the tool aspect by dark dots on a white background for the relation of perpendicularity and horizontal stripes for the relation of parallelism. As an example, Figure 10.9 shows the tables produced for three French textbooks on the market in 2020: *Nouveau Cap Maths CM1* (Charnay and al 2020a), *Méthode*

---

6 https://eduscol.education.fr/2263/organisation-du-temps-scolaire-dans-le-premier-degre, visited November 18, 2021.

Heuristique en Mathématiques (MHM) CM1 (Pinel 2019) and *Maths explicites CM1* (Castioni and al 2020). It should also be noted that this comparison of synoptic tables illustrates the disparity of teaching proposals for geometry in general (the time devoted to it, the relative place of the different sectors and the themes of study, as well as their distribution over the year), for the relations of perpendicularity and parallelism in particular.

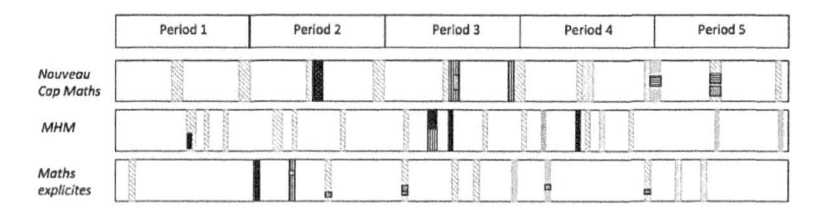

Figure 10.9. *Synoptic tables of the year-long programming in the geometric domain for three textbooks*

At the global level, the criterion for institutional conformity is verified about programming when the learning covered in the textbook corresponds to that which must be taught at the specified level, as identified in the official instructions. Therefore, the criterion is not verified if the programming chosen is not in line with these instructions, that is, if the learning covered is not part of the curriculum for the specified level and/or if learning that is part of the curriculum for the specified level is not covered.

At the global level, the program is educationally adequate when the organization of knowledge corresponds with the intentions declared by the authors of the textbook.

At the local level, the progression is relevant to the teaching of knowledge when the notions of perpendicularity and parallelism are effectively linked to each other and/or to other geometric notions, and when they are approached through their tool aspect within the subsequent geometry sessions (highlighted in Figure 10.9 for the *Nouveau Cap Maths* textbook). On the other hand, when they are not revisited or linked to each other or to other concepts, the progression is considered not relevant.

In a back and forth between the local and the specific, the organization of knowledge is coherent when the concepts and techniques used are, upstream, the object of learning, or when the explicit knowledge is revisited. It is not coherent when concepts or techniques are used without prior learning, or when the meanings of the relations are outside their field of validity (e.g. when the transposition from the plane to the space of the signification of parallelism, "lines that never intersect", is proposed by the textbook), or when the explicit knowledge is never revisited.

10.3.6. Summary

We summarize the different points of attention upon which our analyses are focused by comparing them to the five criteria (Figure 10.10).

	Institutional Conformity	Educational Adequacy declared/proposed	Didactic Quality		
			Relevance	Validity	Coherence
Tasks and types of tasks	Conformity of the proposed types of tasks to the institutional documents	Adequacy of the proposed tasks with the declared intentions of the authors	Relevance of the proposed tasks to the teaching of the relation		Coherence between the proposed tasks and the institutionalized techniques
Techniques	Conformity of the presented techniques to the institutional documents	Adequacy of the proposed techniques with the declared intentions	Relevance of the presentation of the techniques		Coherence between the significations of the relations and the proposed techniques
Knowledge	Conformity with institutional documents of the approached meanings	Adequacy of the introduction of the knowledge with the declared intentions of the authors	Relevance of the teaching for the different significations approached	Mathematical validity of the significations approached	Coherence between the first encounter and the institutionalized knowledge
Ostensives (objects and instruments, languages)	Conformity of the proposed symbols and mathematical notations to the institutional documents	Adequacy in terms of the position of the proposed ostensives with the declared intentions of the authors	Relevance of the choice of objects to the teaching of the relation	Mathematical validity of the use of the instruments	
				Mathematical validity of the symbols and notations	
			Relevance of the language formulations in relation to the teaching of knowledge	Validity of language formulations	
Organizational and planning elements	Conformity with the institutional documents of the program	Adequacy of the programming with the declared intentions of the authors	Relevance of the progression in relation to the teaching of knowledge		Coherence in terms of the organization of knowledge

Figure 10.10. *Summary of the different analysis points*

10.4. Conclusion

Our textbook analysis tool is based on the study of the mathematical and didactic organizations presented in the textbooks. To evaluate the didactic quality of a textbook, we used a previously defined mathematical organization of reference. We thus focused our attention on the types of tasks and techniques proposed, on the associated knowledge at stake, on the ostensives used and on the organizational and planning elements. Relevance of the teaching of the proposed knowledge, validity of the proposals in relation to the mathematical knowledge and coherence in relation to the knowledge being taught are the three criteria that we have retained in order to determine the didactic quality of a textbook.

We have supplemented this analysis with two additional criteria: institutional conformity and educational adequacy between the "declared" and the "proposed". The analysis grid tested on different textbooks, with a view to a comparison to study proposals concerning the teaching of perpendicularity and parallelism relations, seems to us to be unreservedly practical (Guille-Biel Winder and Petitfour 2022). However, to provide this information requires an in-depth didactic study of the textbook, which is not necessarily the responsibility of the teacher. However, we believe that schoolteachers can read summaries in the form of synoptic tables, as illustrated in Figures 10.9 and 10.10, based on analyses carried out by researchers. Our analysis tool can therefore be used to objectivize an evaluation of textbooks relating to the teaching of perpendicularity and parallelism relations in the fourth grade, from which users (mainly teachers) can make an informed decision. This evaluation could also be used by textbook authors to improve the didactic quality of their products. It should also be noted that this grid was designed around the specific study of a geometric theme. The grid filled in for one textbook therefore sheds light on the teaching of geometry in that textbook, but it remains to be seen to what extent it could be adapted to other themes, in order to study the didactic quality of a complete textbook.

Furthermore, we hypothesize that a textbook of "good" didactic quality can contribute to improving the teaching of mathematics at school and promote the learning of pupils. We know, however, that this cannot be sufficient, as teachers' uses of the textbook are dependent on their practices (Arditi 2011) as well as on their "modes of engagement" with the textbook (Remillard 2010).

10.5. References

Arditi, S. (2011). Variabilité des pratiques effectives des professeurs des écoles utilisant un même manuel écrit par des didacticiens. PhD Thesis, Université Paris Diderot.

Bartolini Bussi, M.-G. and Mariotti, M.-A. (2008). Semiotic mediation in the mathematics classroom: Artifacts and signs after a Vygotskian perspective. In *Handbook of International Research in Mathematics Education*, 2nd revised edition, English, L., Bartolini Bussi, M., Jones, G., Lesh, R., Tirosh, D. (eds). Lawrence Erlbaum, Mahwah, NJ.

Bosch, M. and Chevallard, Y. (1999). La sensibilité de l'activité mathématique aux ostensifs. Objet d'étude et problématique. *Recherches en didactique des mathématiques*, 19(1), 77–124.

Bosch, M. and Gascòn, J. (2005). La praxéologie comme unité d'analyse des processus didactiques. In *Balises en didactique des mathématiques*, Mercier, A. and Margolinas, C. (eds). La Pensée Sauvage, Grenoble.

Bourreau, S., Gaspard, P., Graff, O., Rzanny, F. (2016). *J'aime les maths CM1*. Belin, Paris.

Bruner, J. (1960). *The Process of Education*. Harvard University Press, Cambridge.

Castioni, L., Amiot Desfontaine, M., Budon Dubarry, H. (2020). *Maths explicites CM1*. Hachette Education, Paris.

Charnay, R., Anselmo, B., Combier, G., Dussuc, M.-P., Madier, D. (2020a). *Nouveau Cap Maths CM1. Livre-élève*. Hatier, Paris.

Charnay, R., Anselmo, B., Combier, G., Dussuc, M.-P., Madier, D. (2020b). *Nouveau CapMaths CM1. Guide de l'enseignant*. Hatier, Paris.

Chevallard, Y. (1999). Analyse des pratiques enseignantes et didactique des mathématiques : l'approche anthropologique. *Recherches en didactique des mathématiques*, 19(3), 221–266.

Chevallard Y. (2002). Organiser l'étude 3. Écologie & régulation. In *Actes de la 11e école d'été de didactique des mathématiques* (Corps, 21-30 August 2001), Dorier, J.L., Artaud, M., Artigue, M., Berthelot, R., Floris, R. (eds). La Pensée Sauvage, Grenoble.

Chevallard, Y. (2011). La notion d'ingénierie didactique, un concept à refonder. Questionnement et éléments de réponse à partir de la TAD. *Actes de la 15e école d'été de didactique des mathématiques* (Clermont-Ferrand, 16–23 August 2009) [Online]. Available at: http://yves.chevallard.free.fr/spip/spip/IMG/pdf/Cours_de_YC_a_l_EE_2009-2.pdf [Accessed 4 August 2021].

Douady, R. (1986). Jeux de cadres et dialectique outil-objet. *Recherches en didactique des mathématiques*, 7(2), 5–31.

Duplay, J.-P., Demagny, C., Demagny, J.-P., Dias, T. (2008). *La tribu des Maths CM1*. Magnard, Paris.

Dussuc, M.-P., Gerdil-Margueron, G., Mante, M. (2006). Parallélisme au middle school. In *Actes du 32e colloque COPIRELEM* (30–31 May and 1 June 2005, Strasbourg), Rauscher, J.C. (ed.). IREM de Strasbourg.

ERMEL (2006). *Apprentissages géométriques et résolution de problèmes*. Éditions Hatier, Paris.

Guille-Biel Winder, C. and Petitfour, E. (2018). L'enseignement des notions de perpendicularité et de parallélisme dans le manuel Méthode de Singapore. *Grand N*, 102, 540.

Guille-Biel Winder, C. and Petitfour, E. (2019). Enseignement-apprentissage des notions de perpendicularité et de parallélisme en FOURTH-GRADE : que proposent les manuels ? In *Actes du 45e colloque COPIRELEM* (12–14 June 2018), Petitfour, E. (ed.). ARPEME, Paris.

Guille-Biel Winder, C. and Petitfour, E. (2021). Contribution à l'analyse didactique de manuels scolaires numériques du premier degré : une étude de cas. *Education et didactique*, 21(2), 49–76.

Guille-Biel Winder, C. and Petitfour, E. (2022). Mise en fonctionnement d'une grille pour l'analyse didactique de l'enseignement de la géométrie dans des manuels scolaires. *Grand N*, 110, 19–45.

Houdement, C. (2007). À la recherche d'une cohérence entre géométrie de l'école et géométrie du collège. *Repères IREM*, 67, 69–84.

Laparra, M. and Margolinas, C. (2016). *Les premiers apprentissages à la loupe*. De Boeck, Brussels.

MEN (2015). Programme d'enseignement du cycle de consolidation (middle school). *Bulletin Officiel Spécial no. 11 du 26 novembre 2015*. Ministère de l'éducation nationale, France.

MEN (2018a). Modification des programmes d'enseignement du cycle de consolidation. *BOEN no. 30 du 26 juillet 2018*. Ministère de l'éducation nationale, France.

MEN (2018b). Espace et géométrie au middle school. Ministère de l'éducation nationale, France [Online]. Available at: https://eduscol.education.fr/cid101461/ressources-maths-cycle-3.html [Accessed 10 May 2021].

MEN (2019). Attendus de fin d'année et repères annuels de progression du cycle de consolidation. BOEN no. 22 du 29 mai 2019. Ministère de l'éducation nationale, France.

Mounier, E. and Priolet, M. (2015). *Conférence de consensus Nombres et opérations, premiers apprentissages à l'école primaire : les manuels scolaires de mathématiques à l'école primaire – de l'analyse descriptive de l'offre éditoriale à son utilisation en classe élémentaire*. CNESCO, Paris, and IFÉ-ENS, Lyon [Online]. Available at: http://www.cnesco.fr/wp-content/uploads/2015/11/Manuels.pdf [Accessed 12 April 2019].

Petitfour, E. (2017). Outils théoriques d'analyse de l'action instrumentée, au service de l'étude de difficultés d'élèves dyspraxiques en géométrie. *Recherches en didactique des mathématiques*, 37(3–2), 247–288.

Petitfour, E. and Barrier, T. (2019). D'un cadre d'analyse de l'action instrumentée en géométrie à l'élaboration d'un dispositif de travail en dyade au middle school. In *Actes de la 19e école d'été de didactique des mathématiques ARDM*, Coppé, S. and Roditi, E. (eds). La Pensée Sauvage, Grenoble.

Piaget, J. (1964). *Six études de psychologie*. Gonthier-Denoël, Paris.

Pinel, N. (2019). La méthode heuristique de mathématiques. Guide des séances CM1-CM2. [Online]. Available at: https://methodeheuristique.com/modules/modules-cm [Accessed 20 juin 2021].

Rabardel, P. (1995). *Les hommes et les technologies. Approche cognitive des instruments contemporains*. Armand Colin, Paris.

Remillard, J.T. (2010). Modes d'engagement : comprendre les transactions des professeurs avec les ressources curriculaires en mathématiques. In *Ressources vives : le travail documentaire des professeurs en mathématiques*, Gueudet, G. and Trouche, L. (eds). Presses Universitaires de Rennes.

Reymonet, C. (2004). Un cadre expérimental pour l'étude de la géométrie au middle school: Le cas du parallélisme. *Grand N*, 73, 33–48.

Rosenshine, B. (1986). Synthesis of research on explicit teaching. *Educational Leadership*, 43, 60–69 [Online]. Available at: http://formapex.com/telechargementpublic/rosenshine1986c.pdf [Accessed 25 June 2021].

Rouche, N. (2008). *Géométrie. Du quotidien aux mathématiques*. Ellipses, Paris.

Tek Hong, K. (2009). *Manuel de mathématiques CM1 – Cours/cahier d'exercices A – Méthode de Singapour*. La Librairie des Écoles, Paris.

Villani, C. and Torossian, C. (2018). Rapport : 21 mesures pour l'enseignement des mathématiques [Online]. Available at: http://cache.media.education.gouv.fr/file/Fevrier/19/0/Rapport_Villani_Torossian_21_mesures_pour_enseignement_des_mathematiques_8 96190.pdf [Accessed 30 January 2019].

PART 3

Teaching Practices and Training Issues

11

Study on Teacher Appropriation of a Geometry Education Resource

11.1. Introduction

In the early 2000s, researchers and trainers at the IUFM Nord Pas de Calais developed an approach to geometry characterized by the hypothesis for the possibility of a progression adapted to the cognitive development of students (Perrin-Glorian and Godin 2014; Mathé et al. 2020). Learning situations have been produced within this framework, but their implementation in the classroom by teachers uninitiated with the approach has raised new issues. Since 2014, we have been questioning the conditions for the dissemination of situations based on the research of ordinary teaching within the framework of the LéA[1] "réseau de circonscriptions de l'Académie de Lille" (network of districts in the Lille Academy). Relying on a multi-field team of researchers, educational advisors, expert trainers and teachers, we have produced a resource[2] for teaching geometry to the 9- to 12-year-old age group.

We have also organized training sessions, set up coaching and, on the research side, questioned the appropriation by teachers of situations based on this approach. In this text, we report on the follow-up of three teachers, and analyze their practices

---

Chapter written by Christine MANGIANTE-ORSOLA.
1 Place of education associated with Ifé (Institut français de l'éducation).
2 This resource, presented as a website, brings together different types of documents: a general presentation of the approach and its link with the programs, the description of activities to be used in class, student documents, aids and tools for implementation, examples of written traces, etc.

during the implementation of a learning situation designed to help teachers learn about this approach to geometry.

This work is part of a larger questioning on the effects of a training device, and more precisely on the possibility of enriching the practices of trainees, in the more or less long term. Elements from our previous research, based on the observation of the practices of school teachers teaching mathematics, led us to hypothesize that in order to be effective, a training course, regardless of when it is given, must meet at least two conditions: resonate with the immediate (short- and medium-term) professional concerns of the trainee, and be sufficiently holistic and coherent so as to question the coherence of the trainee's own practices (Butlen et al. 2017).

In what follows, we question this hypothesis in the specific framework of a training system based on the LéA project, paying particular attention to the role of geometric knowledge in the process of appropriation by teachers for a situation proposed by the resource. We will start by presenting the context of the research, then specify our questioning and the methodology chosen to study the adaptability of one of the situations designed for the conditions of ordinary teaching, then expose the main elements of analysis resulting from the follow-up with the three teachers, and finally discuss the conditions for enriching their teaching practices in light of the results obtained.

11.2. Research background

11.2.1. *Study on dissemination possibilities in ordinary education*

The situations produced at the beginning of the year 2000 by the researchers and trainers of the IUFM Nord Pas de Calais are the fruit of experiments carried out in the classes of teacher trainers associated with the project, who take an active role in the work of designing these situations.

However, the few attempts to implement these situations in ordinary classes[3] have led us to the following observation: these situations are certainly very rich from the point of view of potential learning for the students, but they are far away from ordinary practices[4] (Leclercq and Mangiante-Orsola 2014). The question of their

---

3 The adjective "ordinary" should not be taken in a pejorative sense.
4 Throughout the text, we use the term "practices" as defined by Robert and Rogalski (2002). Practices refer to "everything the teacher does before, during and after class (conceptions activated when preparing for sessions, various types of knowledge, mathematical and non-mathematical discourse during class, specific gestures, correction of students' productions, etc.)" (*translated by author*).

appropriation by teachers, uninitiated to this approach, was then quickly raised and motivated the choice to initiate a research project aimed at studying the conditions for the diffusion of these situations within ordinary teaching practices.

In September 2014, in the Valenciennes-Denain district, we formed a team composed of two researchers, a National Education Inspector (NEI), three district pedagogical advisors and three expert teacher trainers[5]. During the first two years, this team worked with about 15 teachers from the district who had received in-service training and who volunteered to participate in the experimentation. This design work brought together all the participants around a common project, the elaboration of various documents[6], to be organized and uploaded online, within the framework of a training system. In accordance with our initial desire to question the conditions for the diffusion of the designed situations in ordinary teaching, we have chosen to extend our work of resource production and testing to other districts of the Lille academy, within the framework of training and/or support systems.

From a research point of view, our objective was twofold: on the one hand, to continue to study the possibility of an approach to geometry for elementary school, adapted to the cognitive development of students and allowing for the development of concepts upon which theoretical geometry can be based on at the junior high school level, and, on the other hand, to study the conditions for the dissemination of this approach within ordinary teaching, and the enrichment of teaching practices that this may induce (Mangiante-Orsola and Perrin-Glorian 2016a, 2016b).

11.2.2. *Resource design approach*

Our approach to resource design was that of didactic engineering for resource development and training (Perrin-Glorian 2011). Like all engineering, the didactic engineering resource development aims at producing teaching situations; however, it is supposed to take into account the possibilities for the adaptation to the ordinary conditions of teaching, as part of the design work. In particular, it takes as a core object of study the question regarding the diffusion of these situations within ordinary teaching practices. Therefore, this approach proposes to think of the relationship between research and teaching, not in a top down way in the sense of a transmission from research to teaching, but as an adaptation to ordinary practices, a more dialectical approach that sits between these two poles, nevertheless remaining within the framework of didactic engineering (Brousseau 2013).

---

5 The researcher trainers are (or have been) among the field trainers; some are classroom-based and others not.

6 Preparation sheets, documents for the classroom, materials to cut out, implementation resources, progress markers, presentation texts, etc.

Didactic engineering resource development explicitly assumes two levels of questioning in an attempt to take more account of ordinary teaching practices. The first level of questioning aims at testing the theoretical validity of the situations, that is, their capacity to produce the expected knowledge and to identify the fundamental choices for the engineering. The second level aims at studying the adaptability of the situations for ordinary teaching practices, their real-world conditions in relation to the teachers' practices and the evolution in their perspectives. These two levels are not chronological: when working on the first level, the researcher tries to anticipate the second level, and when working on the second level, they are led to question the first level[7].

Adapting this approach to our project, we designed, within the small group of researchers and trainers, a first situation respecting the hypotheses resulting from the research. Then, the teachers associated with the project tested it in class. Overlapping our different "points of view", we analyzed together different implementations, examined and adjusted the situation so as to propose it to other teachers, tested it again, and so on. We thus produced this resource (i.e. the situation presented, the detailed description of its development, the implementation advice, etc.) according to iterative loops. As the experimentation progressed, we retested it again and again with a growing sample of teachers.

We first tested it with teachers from the Valenciennes-Denain district, and then with teachers from other districts who received lighter support.

Therefore, throughout our project, we were led to focus on two levels of questioning in a dialectical manner. It is on the second level of questioning that the rest of this text focuses. However, before explaining our research approach, we need to present the working hypotheses on which the resource produced within the LéA is based.

11.2.3. *A working methodology based on assumptions*

11.2.3.1. *Assumptions about pupils' learning*

The approach developed in the Nord Pas de Calais region is in line with the work of Duval (2005) and Duval and Godin (2006). Taking up the idea that the students'

---

[7] For example, following requests from teachers, we have been led to propose possible adaptations of the situation according to the level of the students (management of heterogeneity). Responding to this request is a matter for the second level (because it contributes to taking into account the constraints of ordinary teaching), but the response to be given must be thought out by reexamining the first level (because the adaptations to be planned must not distort the situation).

relationship to figures is one of the key points for their entry into geometry, Perrin-Glorian and Godin distinguish three visions of figures (Figure 11.1), according to these viewpoints that we are able to bring to them (Perrin-Glorian and Godin 2014):

– When a "surfaces" vision is activated, the figure is seen as a collection of surfaces that are juxtaposed or overlapping. In a "surfaces" vision, lines and points may appear, but they are the edges of surfaces, the vertices of surfaces or the intersections of surface edges. Therefore, in a "surfaces" vision, lines and points are objects that are subject to surfaces.

– When a "lines" vision is activated, the figure is seen as a network of lines that have their own existence. Points may appear but they are the ends of lines or intersections of lines already drawn. Lines can be created from surface contours. Therefore, in a "line" vision, the points are subject to lines.

– When we activate a "points" vision, we see this figure as a configuration of points that have their own existence. Points can be created by the intersection of two lines that are drawn for this purpose, and points can define lines.

Let us specify that each vision includes the previous one, so that in the long run, what is aimed at is the acquisition of a certain flexibility of the glance, authorizing the individual to activate one or the other of these complementary visions.

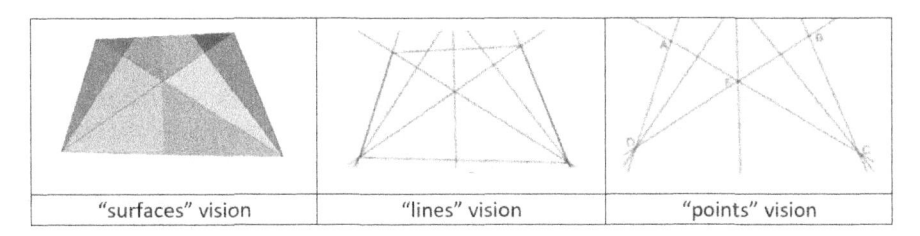

Figure 11.1. *Three different visions of the same figure*

11.2.3.2. *Assumptions about teaching geometry*

Based on this analysis of "surfaces", "lines" and "points" visions, the approach developed in the Nord Pas de Calais region is based on the hypothesis that certain situations would accompany the pupils' change in viewpoint towards figures, leading them to progressively enrich their "surfaces" vision, with a "lines" vision, and then a "points" vision. These situations are *figure restoration* situations.

In a restoration situation, a model figure is provided (full size or not), and the students have a part of the figure to obtain (starting figure) as well as various instruments. Either the starting figure, or some of the instruments allow for the transfer of the 2-dimensional (2D) information of the initial figure, without giving all the information, so that the student can start the restoration with a "surface" vision of the figures (Figure 11.2). The starting figure and the instruments available to the students are didactic variables whose choice of values has a consequent impact on the procedures expected of the students, which allows the teacher to accompany their change in their apprehension of figures.

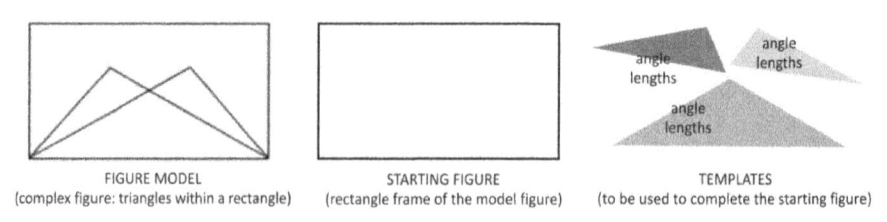

FIGURE MODEL
(complex figure: triangles within a rectangle)

STARTING FIGURE
(rectangle frame of the model figure)

TEMPLATES
(to be used to complete the starting figure)

Figure 11.2. *Figure restoration situations*

11.2.3.3. *Assumptions about training*

Prior to the work carried out with the LéA, we conducted observations of geometry sessions and made certain observations. The teaching of geometry is generally centered on the acquisition of vocabulary and the mastery of typical instruments, and is mainly based on ostensive practices. The concepts of the line, parallel or perpendicular lines, are taught without many links to the reproduction of figures; there is little or no explicit work on the transfer of quantities independent of their measurement (Mangiante-Orsola and Perrin-Glorian 2016b). Regarding the situations produced, within the framework of the approach to which we refer, we were able to observe that the teachers had difficulties in linking these to the learning issues of geometry teaching, which can be a hindrance to the enrichment of their practices, beyond the simple implementation of these teaching situations. Moreover, playing on the didactic variables and taking them into account at the time of implementation (especially during the tasks involving instruments) is not self-evident. This is why we have retained, for the continuation of our work, the hypothesis according to which it is necessary to think of a resource which proposes situations allowing both the pupils to enrich their viewpoint towards figures and those that allow teachers to train themselves through the implementation of these situations. Therefore, via the joint observation of implementations, we could lead teachers to question their practices at different levels of practice organization (Butlen et al. 2017): at a global level (the issues of geometry teaching), the local level (the implementation of a session) or the micro level (such as the answer given to a particular student's question). Moreover, we felt that we needed to "think about a certain 'continuity' in the enrichment of practices, that is, to think about the possibilities of a progressive integration of new practices within existing practices" (Leclercq and Mangiante-Orsola 2014, *translated by author*). We therefore designed a situation that could lead teachers to question their practices without destabilizing them too much. Here is the path followed to design this situation "to enter the approach".

11.2.4. Designing a situation using the didactic engineering approach for development

Investing the first level of questioning of didactic engineering resource development, we decide to design a restoration situation based on the model figure (Figure 11.3). A quick analysis shows that this figure consists of a quadrilateral inside of which there are two smaller triangles (which we will denote as T1 and T2), or three triangles, or two large triangles that overlap. We can also note that some sides of the T1 and T2 triangles follow the diagonals of the quadrilateral. There are no right angles.

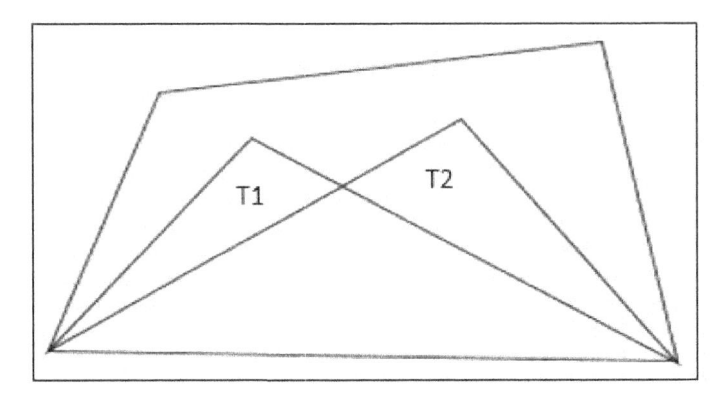

Figure 11.3. *Model figure*

Based on the theoretical tools of epistemological and cognitive analysis, and the theory of didactic situations, we draw from this first analysis the potential learning. Therefore, depending on the approach and the instruments chosen, restoring this figure could enable the pupils to carry out decompositions into simple figures, to carry out assemblies by juxtaposition or by overlapping triangles, to take into account alignments of segments (alignment of the sides of the T1 and T2 triangles), and alignments of segments and points (alignment of the sides of T1 and T2 triangles with the vertices of the quadrilateral).

Investigating the second level, we decide to facilitate the appropriation of our approach by the teachers through a situation (which we call "triangles on a quadrilateral") organized according to several phases, during which the same task is expected from the students (to restore the model figure), but which differs through a play of the didactic variables involved (starting figure and instruments).

In order to avoid difficulties that could arise during the implementation of this situation, we have designed preliminary activities during which students will be initiated into the use of templates, which can be juxtaposed or overlapped, used to extend lines, to leave visible construction lines, to justify their procedures, etc.

In addition, we have included in the resource different types of documents for teachers, which facilitate the implementation (student documents, individual material, collective material, contextualized and decontextualized implementation aids, elements to be institutionalized, etc.).

11.3. Focus on the adaptability of this situation to ordinary education

11.3.1. *Details about the theoretical framework and the research question*

To question the adaptability of the situation thus conceived, it is necessary to carry out numerous observations of effective implementation in the classroom, and to interpret these observations in light of a study on the normal practices of the teachers under monitoring.

In this perspective, we place our work within the framework of the "dual didactic and ergonomic approach to mathematics education practices" (Robert and Rogalski 2002, *translated by author*) which considers the teacher as an individual in a work situation, subject to specific constraints of the profession. According to the work developed within this framework, teaching practices constitute a complex, coherent and stable system, and this system is put in place from the first years of practice. In fact, elements present from the initial training onwards can mark the ways in which the first teaching experiences are sedimented, and be at the origin of the introduction of a certain set of practices (Mangiante-Orsola 2012). Moreover, the beginning practices of teachers quickly become their norm (Butlen 2005). Of course, this does not mean that everything is set in stone at the end of initial training. The work carried out within the framework of the dual approach attests to the possibility for each teacher to invest a certain "margin for manoeuvre". Therefore, teaching practices constitute a complex, stable and coherent system that could, under certain conditions, evolve over time and through experience.

According to our work, each teacher has a certain potential repertoire of practices, and the role of training is to enrich this repertoire, that is, to increase the range of options available to the teacher (Mangiante-Orsola 2012). In this regard, it should be noted that offering teachers alternatives is not enough to enrich their practices. Training must propose alternatives that will ultimately promote learning.

Wishing to question the adaptability of the situations conceived within the framework of the LeA, we wonder about the factors likely to favor the enrichment of these practices. More precisely, we have chosen to study the practices of teachers who use the situation "to enter the process", in order to detect possible traces of resonance between the proposals made via the resource and their normal practices, and then to look for signs for potential enrichment.

11.3.2. *Presentation on the follow-up of teachers, details of the research question and the methodology*

To answer these questions, we chose to follow three teachers in the context of an in-service training program involving nine Grade 4 to Grade 6 teachers, during which training sessions alternated with classroom experimentation sessions. During the first training session, the main theoretical elements[8] were presented to the trainees. An evaluation and "preliminary activities" to be implemented in the classroom are then made available to them. During the second session, the situation to "enter the process" is presented to them and is the subject of the main experiment to be carried out. Then, a third and final session allows them to share their experiences, report on difficulties encountered and consider possible extensions.

The three teachers selected were given individualized support, since interviews followed each of the sessions observed, which enabled them to enrich their analysis[9]. All three are experienced teachers and work in the same priority education network. Valérie and Fabien work in the same school and are each responsible for a Grade 4/Grade 5 class. Valérie is the headmistress. Fabien is preparing for an exam to become a teacher. Céline works in another school, in a Grade 4 class.

The monitoring system setup allows us to collect the data necessary for our research: for each of the three teachers, we have the video recording of a preliminary session before the beginning of the training, followed by four sessions corresponding to the four phases of the "triangles on a quadrilateral" situation, and the audio recording of the interviews conducted "on the spot" after each session. The data thus collected are characterized by common parameters. Indeed, when analyzing teaching practices, many factors can intervene, but here, the teachers work in the same district, or even the same school for two of them, and in the same REP+

---

[8] The challenges of teaching geometry, the way students look at figures, the restoration of figures, the use of instruments, etc., were discussed.

[9] However, no additional documentation was provided to them during these interviews.

zone[10]. None of them is a beginner. Moreover, these three teachers implemented the same situation from the same document, and received the same training and support.

Setting so many parameters should allow us to highlight the particular choices of each teacher, and facilitate the uncovering of intra-individual regularities revealing the teacher's priorities and the coherence throughout their practice. We hope to then be able to detect possible resonances between these regularities and the choices characteristic of the situation and the resource (in terms of both content and form). Are these resonances, between the choices made by the authors of the resource and the coherence of the trainee practices, sufficient for the teachers to appropriate this situation? Are certain elements taken up in the same way? Are others added or rejected? For what reason(s)? What is the role of the geometric knowledge involved in the origin of the discrepancies observed between the expectations of the designers of the resource and the observed implementations? Figure 11.4 illustrates our questioning and the analytical approach that we will now develop.

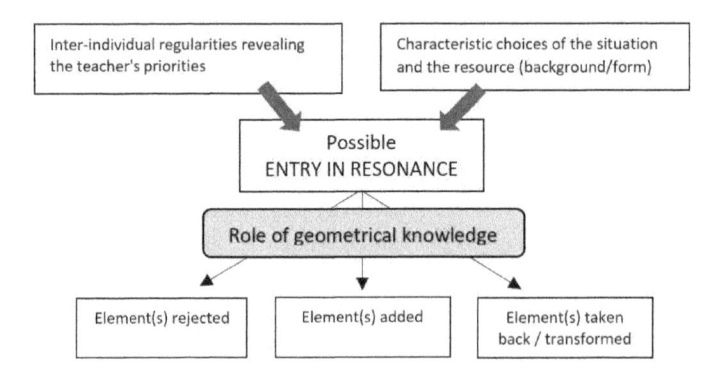

Figure 11.4. *Questioning and methodology implemented*

11.3.3. *Presentation of the analysis methodology*

In order to carry out this study, we begin by carrying out an a priori analysis of the situation; next, we make an analysis of the teacher's activity by anticipation (for this, we focus on identifying the knowledge to be mobilized by the teacher during the implementation of the session, the professional gestures to be implemented and the difficulties that they might encounter). Then, for each of the three teachers, we rely on the analysis of the five sessions observed, to identify individual regularities

---

10 In France, REP is a "priority support network" of schools that allows schools located in a disadvantaged social environment to benefit from specific facilities such as smaller class groups.

likely to characterize the coherence of their practices. In order to do this, we rely on the discrepancies detected between the expectations of the authors of the resource and the actual implementation, as well as on the a priori analysis of the situation, completed by the anticipatory analysis of the teacher's activity. Next, we compare these individual regularities, the characteristic choices of the situation presented and those of the resource, in order to reveal possible resonances.

Finally, we study some aspects of the implementation, in relation with this resonance, in order to interpret the origin of the discrepancies observed with regard to the geometric knowledge at stake.

11.4. Elements of the analysis

11.4.1. *Analysis a priori of the situation and anticipatory analysis of the teacher's activity*

11.4.1.1. *Key elements of the analysis a priori*

The sequence is organized in four successive phases. Although the instruction given is always the same: "restore the model figure from the given starting figure using the instruments available", the knowledge to be used is not the same from one phase to the next, so that the pupils' view of the figure can thus be progressively enriched (Figure 11.5).

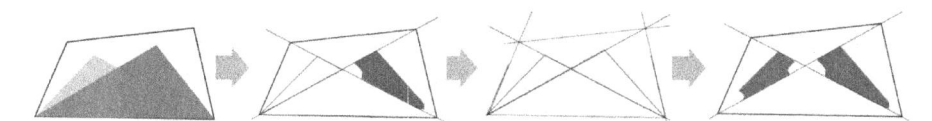

Figure 11.5. *A sequence organized into four phases. For a color version of this figure, see www.iste.co.uk/guille/tangible.zip*

In the first phase (Appendix 4), the students restore the figure from the templates of the two large triangles. The quadrilateral frame is given (this is the starting figure). They are thus led to move from an analysis of the figure in terms of an assembly of juxtaposed surfaces to an analysis in terms of overlapping surfaces, to move from a vision of solid surfaces to a vision of surface contours. The alignment of the sides of the small triangles is taken care of by the material (some of the sides of the large triangles are constituted by the alignment of two sides of the T1 and T2 triangles). The students do not have to use it explicitly, but they are led to apprehend it through the gestures made, since they will have to trace all or part of the contour for each of the two templates.

In the second phase, a different approach is presented. The students have to extend the sides of the T1 triangle and thus explicitly use the segment alignments (the sides of the small triangles), in order to be able to place the template and extend the sides of the T2 triangle in order to complete the outline. Note that it is a question here of passing from a "surface" vision to a "line" vision, since the pupil must employ the outline of the triangle to make a network of segments appear.

In the third phase, students are given a starting figure and a non-scale ruler. The model is at a reduced size in order to avoid the transfer of lengths. The students extend the two sides of the quadrilateral that have been partially traced, thus revealing the straight lines that support these two sides. Then, in order to draw the missing side, they must determine two points. These two points, which are the missing vertices of the quadrilateral, can be obtained by the intersection of two lines supporting a side of the quadrilateral and a diagonal. Once the missing vertices are obtained, they can now draw the fourth side. Note that this involves activating a "points" vision since the student must create points by intersecting two lines with the intention of determining another line.

In the fourth phase, the choice of the starting figure and the instruments leads the students to revisit this knowledge by first tracing the diagonals of the quadrilateral in order to position the two templates.

11.4.1.2. *Key elements of the anticipatory analysis*

Our analysis of the teacher's activity is the result of numerous cross-observations carried out in the classes (let us specify, for example, that during the first year of work within the LéA, 15 classes were observed). It focuses on different levels of practice.

At a global level, we retain from these first observations, in particular, the need for the teacher to take into account the challenges of teaching geometry, and to set up a certain didactic contract[11] in accordance with these challenges. At a local level, we point out the need for a certain a priori analysis of the procedures expected from the students (these procedures are described in the resource), for the observation and explanation of these procedures with a view to the institutionalization of knowledge (text of knowledge specified in the resource), and the importance attributed to language. We also note the importance of helping students by getting them to explain "what is causing them problems". This is why, at a more micro level, we pay attention to certain specificities related to each of the four phases that we cannot describe here in their entirety, but for which we present the key elements. In phase 1, the setting up of the didactic contract (overlapping of the authorized templates) can

---

11 Among other things, we can examine the teacher's discourse on the attention needing to be paid to the precision of tracings or the correctness of the procedures involved.

constitute an obstacle for students, in which case it is up to the teacher to authorize them. The alignment carried by certain edges of the templates should be emphasized, for example, by asking the students to highlight them with a color.

In phase 2, the teacher should be aware that the student is confronted with the need to trace elements in order to place the template and, in doing so, be led to use the alignments, noted in the previous phase. In phase 3, the main obstacle is related to the need to identify two points in order to draw the missing side. However, identifying these points is not easy. In fact, they appear on the model figure as the intersection of three straight lines (diagonal and straight lines, supporting two sides of the quadrilateral), and on the starting figure, as the intersection of two straight lines (supporting sides of triangles and one side of the quadrilateral). However, many students who have obtained the points by extending the lines on the starting figure are not aware that these are precisely the points they are looking for. The teacher should encourage the students to go back and forth between the starting figure and the model figure, which is essential for this awareness. Finally, in phase 4, the teacher may have to help the students become aware of the need to trace elements in order to place the templates.

11.4.2. *Analysis of practices*

11.4.2.1. *Analysis of Fabien's practices*

Among Fabien's routines identified, let us start with the one that consists of questioning the students at the beginning of the session in order to get them to formulate hypotheses, which he notes on the board and to which he will return later. The routine nature of this organization is reinforced by the use of a ritualized coding of the hypotheses formulated: the teacher notes the students' proposals on the board, adds a few question marks next to those that need to be tested, which they will erase as the class's reflection progresses. The proposal in the resource resonates with this routine. Indeed, it is specified that the teacher could, at the beginning of the session, note down on the board the list of elements identified on the figure by the pupils, by making, for example, two "I observe/I see" columns (implicitly with the aim of ensuring that this devolution time at the beginning of the session initiates the planned institutionalization). Nevertheless, it should be noted that, in connection with the instructions to be given during the first phase, a clarification is provided in the resource: "It is important to present the model figure collectively, but not to make this presentation last too long. Simply note the main elements of the figure identified by the students. This list can be completed as the sequence progresses." However, during phase 1, Fabien lets the students express themselves at length (15 min.). They mentioned the presence of various simple figures: "right-angled triangle", "trapezoid", "rectangle", "axes of symmetry", "triangles that overlap" or even "center of the figure".

A pupil comes to the board, draws additional segments while explaining: "We could add lines to have additional triangles." Let us note here that this pupil mobilizes a "lines" vision (they trace a network of segments from the contours of surfaces). During the research phase, one pupil tries to build on their classmate's proposal, but they are unable to do so because they use a "surface" vision and have not yet sufficiently enriched their vision to use a "line" vision (Figure 11.6). The teacher observes their attempts, asks them to explain what they are doing, but does not intervene further, which is characteristic of their regulation routine: Fabien circulates in the classroom, asks the pupils to explain their procedure, but he does not support the formulations, does not ask questions or provide any particular help. Therefore, Fabien valued the research attitude of his students, but did not allow himself to intervene beyond one or two questions aimed at clarifying the procedure.

Figure 11.6. *A student destabilized by another student's proposal*

During phase 1, he offered some of the more geometrically inclined students' false templates with the intention of getting them to look harder. However, he did not realize that the addition of an outlier template constitutes as a resource and that, consequently, this element of differentiation should be reserved for students who are having difficulty restoring this figure, and more specifically, those who do not take information from the model figure in order to be able to restore it. Indeed, as specified in the resource, "the presence of an outlier template forces one to identify on the model, the templates corresponding to triangles of the model figure and thus to see these triangles". Therefore, it is the teacher's consideration of the need to install a certain didactic contract that is at stake here. Throughout sessions 1 and 2, Fabien remained in the background in order to promote the spontaneous expression of the pupils.

Believing that students learn by sharing with each other, he does not identify elements for the institutionalization of knowledge. Moreover, he does not choose the students who will come to present their procedures during the sharing of knowledge. However, during the interview following phase 2, he explains: "I find myself with a lot of superfluous procedures, I find myself navigating, sometimes useless things." The support provided will allow Fabien to question his normal practices in relation to the management of procedures implemented by the students, and to enrich these. During the following sessions, he will allow himself to solicit the students according to the procedures implemented, with a view to institutionalizing knowledge. Therefore, the didactic knowledge provided during the interviews enabled Fabien to envisage a slightly different way of managing the sharing phase, in order to better take into account the geometric knowledge at stake.

11.4.2.2. Analysis of Céline's practices

The analysis of the five sessions observed reveals very marked regularities in the way Céline organizes her students' work. She frequently divides the students into workshops (which requires a great deal of preparation). While some of the students were working independently, Céline took charge of a workshop, thus reproducing the usual functioning of a kindergarten class. At the end of the workshops, Céline organizes group time for sharing and institutionalizing knowledge.

The students in her REP+ class are not very independent, quite noisy and classroom management is not easy. However, Céline insists on this organization because it allows her to be as close as possible to each student (especially during the implementation of this geometry sequence). Observation of student activity is an important part of her practice. In the resource, for each of the phases, there is a section entitled "Observe students' procedures." For example, in phase 2, it is stated: "Some students do not perceive the alignments and simply place the templates approximately. Others take into account only one alignment." In addition, a document annexed to the description of the sequence lists the procedures observed in the classes associated with the LeA. This choice obviously resonates with Céline's concerns. Moreover, during the first session, she takes notes on the students' procedures, using an observation grid that she designed herself. Note that not only does Céline adhere to the project presented in the resource, but she also enriches it with an additional tool.

Céline does not just observe the students; she systematically solicits them. She asks each student, "Tell me how you did it." She also pays particular attention to the gestures made. She asks: "Show me again how you did it with your ruler," so that she can observe the positioning of the instruments. Céline also sets up a routine leading the students to justify their procedures, which allows her to enrich the exchanges even more. She asks, "How did you do the trace?", "Are you sure?",

"What helps you?", "What did you base it on?", and thus brings out the elements of the figure to be taken into account when tracing. While the resource emphasizes the importance of observing procedures, it does not specify how to intervene with the students. Céline therefore takes the initiative. When we asked her in an interview, "Why make each student explain the procedure?", she replied, "For two reasons: to understand better and to get them to use the vocabulary."

This questioning routine also allows Céline to provide these students with adapted help. In fact, she helps them to identify the missing elements. For example, during phase 2, she intervenes with a student who has extended only one side of the triangle, and therefore does not have enough elements to place the template. Céline slides the template along the traced line to show that its position cannot be determined with precision (see Figure 11.7). She asks, "Where do you place the template on the model figure?" and explains, "To be more precise you're missing something!" She even adds that "you need at least two directions, that is, two straight lines to be able to place the template! Extending one side of the triangle is not enough!". In doing so, she mobilizes geometric knowledge that is not included in the resource: here, a straight line is not enough to place a template. In phase 3, she will do the same thing, explaining, for example, that a point does not provide a direction for a line.

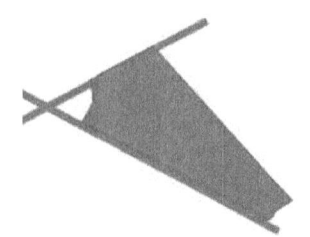

Figure 11.7. *Céline's intervention*

Therefore, our observations show that Céline has truly taken ownership of the process and has a keen understanding of the learning issues. Several initiatives on her part testify to this. Indeed, throughout the sequence, she takes care to install a didactic contract consistent with the approach. During phase 1, she said: "You have analyzed with the material available," and on a collective poster, she wrote: "I must analyze a complex figure with my geometry tools." In phase 3, she tells a student who wants to erase the construction lines, "This is not a drawing that we're doing, this is geometry." The words she uses reveal her understanding of what is at stake in the transition from one to the other. In phase 2, in order to emphasize the need to make each edge of a template coincide with a straight line, she says that the template "hugs" each of the two segments.

Finally, she takes various initiatives as we have already pointed out, and even goes so far as to integrate our proposals into a longer-term teaching project. As early as phase 2, she asks the students who have finished to write at the bottom of their sheet of paper how they did it, and justifies her choice to us, explaining that this makes it possible to manage the heterogeneity of the students – "Some finished before the others"; that it allows her to gather information on the procedures implemented – "I can better understand what they did"; and that it makes it possible to develop an adapted geometric language – "It trains them to write geometry". Because she aims to work on the writing of a construction program, she insists on the chronology of the drawings by saying "You are going to write below all the steps in order." In addition, she produces a display that shows different functions of the dynamic geometry environment software (Geogebra), linking them to elements institutionalized in the figure restoration situation.

Therefore, Céline relies on our proposals to enrich her geometry teaching project, her mastery of mathematics allowing her to make the necessary links between the different sequences planned.

11.4.2.3. Analysis of Valérie's practices

Valérie pays particular attention to language, to the richness of the vocabulary used by the students, and encourages them to consult their dictionary as soon as they encounter a new word. During this sequence, we note (and she will confirm this during the interviews) that she seeks to develop "a certain culture" in her REP+ students. She welcomed our proposals with enthusiasm, presented the figures to be restored to her pupils as if they were "restoring a stained-glass window in a castle", and took the opportunity to ask them about the different possible interpretations of the word "restoration".

Valérie gets her students to work quickly. The classroom climate is conducive to individual research. A characteristic routine of her practices is the regulation process, Valérie does not take global information on the students' activity and, in so doing, deprives herself of interesting procedures to exploit. She intervenes with the students in research, passing from one to another by following the arrangement of the tables, and does not question the students who have succeeded. During phase 1, her attention is focused on the success of the restoration of the figure to the detriment of the analysis of the procedures and the difficulties encountered.

Valérie spends a lot of energy trying to be with everyone, but she does not get students to identify what they are having trouble with. For example, during phase 4, she intervenes with a student who does not know where to place the two templates.

Encouraging the student to express this difficulty could help them realize that the figure is missing elements that would allow the templates to be placed there, in particular straight lines that must coincide with two of their edges. Valérie does not question them. She simply points to the diagonals drawn in red on the display made at the end of the previous session. The student reproduces the diagonals, but as soon as Valérie moves away to look after another student, they partially erase them. Indeed, using a "surface" vision, they try to make the triangle formed by two half-diagonals appear, the one whose one side coincides with the base of the quadrilateral. Then, they place one of the two templates on one of the sides thus traced, but cannot continue because a direction is still missing (Figure 11.8). Because the diagonals were not brought in as "answers" to "his problem", the student cannot grasp the teacher's proposal.

Figure 11.8. *A student's procedure (phase 4)*

It is true that Valérie is more attentive to the precision of the tracings than to the correctness of the procedures. From the very first session, Valérie explains that in geometry, we must not do "the approximate", "not rely on the eye" and that "the tools will help us to be precise". She explains that it is necessary to be "clean", "neat", "not to press too much on the pencil, to extend, to go beyond, etc.", "that we have the right to erase, etc.". She concludes the session by saying: "Geometry is about having something exact." This precision is certainly very important to her. However, in the resource, it is clearly stated: "It must be understood that what is targeted is more the accuracy of the procedures used than the precision of the tracings." During the training, the main elements justifying an approach to geometry without recourse to measurement were presented. In addition, it was explained to the teachers that a distinction should be made between "procedural correctness" and "accuracy of tracing", and that the challenge of teaching geometry was more "procedural correctness", even though it was important to get students to make accurate tracings. In the resource, the adjectives "right" and "precise" are used, but Valérie uses another word, "exact", and we assume that she means "rigorous in terms of length transfers".

We note, however, that the didactic contract set up by Valérie evolves. Indeed, from phase 2, she uses the word "reference" and insists on the students who have clearly perceived her requirement. When she asks: "What was the contract in relation to the construction of a figure? [...] What is the purpose of the template?", one student replied: "To draw lines, to make reference points." Another explained: "I looked at the figure, I took the segment and extended it." Valérie then insists: "What is that called?" The student answers: "A reference." According to Valérie, a "reference" is, for example, the sides of the triangle already traced in S2, the vertices of the quadrilateral in S3, the diagonals in phase 4, etc.

From the analysis of the different sessions, an evolution emerges. During phase 1, the idea is to take references in order to make something that is "exact". Then, little by little, the word "reference" is used to designate the elements of the figure (already traced or to be traced), to succeed in restoring the figure. Therefore, in the first sessions, the graphical purpose was predominant, but little by little, through the use of the word "reference", Valérie was led to designate geometric objects and relationships and, in so doing, prioritizes geometric purpose.

Valérie adheres to our project because it resonates with some of her priorities (interest in the precision of language). Then, little by little, thanks to the coaching, she enriched her regulation routine (observing all the students before intervening with some of them) to finally come to question herself on the stakes of teaching geometry. We can see that some of the didactic knowledge presented during the first training sessions gradually made sense to Valérie as the sessions progressed.

11.5. Conclusion

The LéA team wanted to propose, via a resource, teaching situations likely to provoke sufficient changes in teachers' practices, with regard to the potential learning of the students, but limited enough so as to not destabilize existing practices too much. We cannot attest to an enrichment of these teachers' practices in the long term. However, we can identify implementations that are conducive to potential change.

In spite of a very individualized class management that could suggest a lack of collective time or even a lack of institutionalization, Céline succeeded in bringing out the geometric knowledge at stake and in having her students make it explicit. Not only does she adhere to our project, but she also enriches it with new proposals, thus demonstrating her understanding of the challenges facing the teaching of geometry and her mastery of knowledge. Fabien and Valérie adhere to our proposals for different reasons, but some of their priorities, linked to their representations of geometry teaching or teaching in general, come into conflict with some of the choices of the resource, and their mastery of the geometric and didactic knowledge to be mobilized is sometimes lacking. Nevertheless, each of them will make adjustments to their practices through the support provided. Fabien allowed himself to intervene more in the exchanges, with a view to bringing out the knowledge to be institutionalized. Valérie became aware of the need not to focus on the precision of the tracings, so as to be able to accompany her students in the conceptualization work inherent to geometry.

Therefore, the follow-up with these three teachers led us to note convergences between some of their priorities and certain characteristics of the situation presented in the resource, thus creating resonances likely to encourage their support. Some elements are taken up by the three teachers; others are added (notably by Céline) or rejected (by Valérie and Fabien). The didactic and geometric knowledge mobilized by Céline appeared to facilitate her appropriation of our proposals. However, some of the expected effects on practice seem to be limited in Fabien and Valérie, due to an insufficient mastery of geometric knowledge or a lack of consideration of the issues at stake, or even due to unsuitable routines. Adhering to a proposal made during training does not necessarily lead to an enrichment of practices that potentially promotes learning.

The question as to the adaptability of a situation produced by research to ordinary teaching is complex. The design and experimentation process of the resource, designed according to iterative loops, aims at encouraging this adaptability. However, this adaptability necessarily has its limits, as it depends on the priorities of each person, on their normal practices, as well as on the mobilization of geometric and didactic knowledge. Following this work, we modified our resource, completed the document devoted to implementation advice, and used examples from our classroom observations to better highlight the link between the teacher's choices and the learning issues.

We have also designed a training course[12] aimed at getting teachers to question their practices, based on the analysis of an evaluation proposed to their students. We now plan to further question the role of coherent practices in the process of appropriation by teachers of this situation. In this way, we will be able to continue our testing of the hypothesis that in order to be effective, training, regardless of when it is provided, must meet at least two conditions: it must resonate with the immediate (short- and medium-term) professional concerns of the trainee, and be sufficiently holistic and coherent enough so as to question the coherence of the trainee's own practices (Butlen et al. 2017).

---

12 Via the M@gistère platform (an e-learning platform offered in France to all national education personnel).

11.6. References

Brousseau, G. (2013). Introduction à l'ingénierie didactique [Online]. Available at: guy-brousseau.com/wp-content/uploads/2013/12/introduction-à-lingénierie-didactique3.pdf.

Butlen, D. (2005). Apprentissages mathématiques à l'école élémentaire. Des difficultés des élèves des milieux populaires aux stratégies de formation des professeurs des écoles. Synthesis note, Université de Paris-Diderot, Paris, IREM Paris 7.

Butlen, D., Mangiante-Orsola, C., Masselot, P. (2017). Routines et gestes professionnels, un outil pour l'analyse des pratiques effectives et pour la formation des pratiques des professeurs des écoles en mathématiques. *Recherches en didactiques*, 2017(2), 25–40.

Duval, R. (2005). Les conditions cognitives de l'apprentissage de la géométrie : développement de la visualisation, différenciation des raisonnements et coordination de leurs fonctionnements. *Annales de didactique et de sciences cognitives*, 10, 5–53.

Duval R. and Godin M. (2006). Les changements de regard nécessaires sur les figures. *Grand N*, 76, 7–27.

Leclercq, R. and Mangiante-Orsola, C. (2014). Étude d'un dispositif articulant production de ressources et formation continue en géométrie : quels effets sur les pratiques des enseignants ? *Actes du XLème colloque de la COPIRELEM*. IREM de Nantes.

Mangiante-Orsola, C. (2012). Une étude de la cohérence en germe dans les pratiques de professeurs des écoles en formation initiale puis débutants. *Recherches en didactique des mathématiques*, 32(3), 289–331.

Mangiante-Orsola, C. and Perrin-Glorian, M.-J. (2016a). Elaboration de ressources pour la classe, interface entre recherche et enseignement ordinaire. In *Questionner l'espace. Les méthodes de recherche en didactiques*, Chopin M.P., Cohen-Azria C., Orange-Ravachol D. (eds). Presses Universitaires du Septentrion, Lille.

Mangiante-Orsola, C. and Perrin-Glorian, M.-J. (2016b). Ingénierie didactique de développement en géométrie au cycle 3 dans le cadre du LéA Valenciennes-Denain. *Actes du séminaire national de didactique des mathématiques*. IREM Paris 7, Paris.

Mathé, A.-C., Barrier, T., Perrin-Glorian, M.-J. (2020). *Enseigner la géométrie élémentaire. Enjeux, ruptures et continuités*. Académia L'Harmattann, Louvain-la-Neuve.

Perrin-Glorian, M.-J. (2011). L'ingénierie didactique à l'interface de la recherche avec l'enseignement. Développement de ressources et formation des enseignants. In *En amont et en aval des ingénieries didactiques*, Margolinas, C. et al. (eds). La Pensée Sauvage, Grenoble.

Perrin-Glorian, M.-J. and Godin, M. (2014). De la reproduction de figures géométriques avec des instruments vers leur caractérisation par des énoncés. *Math-école*, 222, 26–36.

Robert, A. and Rogalski, J. (2002). Le système complexe et cohérent des pratiques des enseignants de mathématiques : une double approche. *La revue canadienne de l'enseignement des sciences, des mathématiques et des technologies*, 2(4), 505–528.

12

Geometric Reasoning in Grades 4 to 6, the Teacher's Role: Methodological Overview and Results

12.1. Introduction

This work focuses on the reproduction of figures in plane geometry, in Grades 4–6, which in France includes the last two years of primary school and the first year of secondary school – a topic developed in a thesis by Blanquart (2020). From an institutional point of view, the July 2018 program specifies that the situations proposed to the pupils should lead them to gradually transition from a geometry where objects and their properties are controlled by perception, to a geometry placed under the control of instruments, graduating to a geometry whose validation is based on argumentation and reasoning. The play on the constraints of the situations, on the variations of the mediums at the disposal of the pupils, on the instruments authorized (or not), is at the root of this evolution of procedures and the enrichment of knowledge. In this context, the aim of our research is to identify how, in plane geometry, Grade 4–6 teachers integrate (or not), throughout their teaching project, the valid or erroneous reasoning implemented explicitly or implicitly by the pupils. This chapter reports on the methodology used for this research and the conclusions on the role of teachers

12.2. Theoretical choices and the problem statement

In this framework of the theory of didactical situations (TDS) (Brousseau 1998), an adidactic situation is a situation that can be associated with the teaching of a

---

Chapter written by Sylvie BLANQUART.

knowledge or a skill identified by the teacher, where the intention to teach is eliminated in order to leave as much initiative as possible open to the pupil, and to allow them to act on their own and make decisions (Brousseau 1998). An adidactic situation is always part of a didactic intention (Margolinas 2004; Schneider and Mercier 2008). It leads to a phase of institutionalization during which the teacher makes the pupils recognize the knowledge targeted as an optimal answer to the question initially posed, and more generally to the class of problems from which this question stems (Schneider and Mercier 2008). Managing this institutionalization phase can be complex for the teacher. They have to rely on the reasoning and knowledge that they perceive in the pupils' productions, during different phases, in order to lead the class to a collective construction of the knowledge in question. When the pupils produce a variety of arguments, the teacher has to choose from among them those he will be able to exploit.

12.2.1. *Geometrical paradigms*

Houdement and Kuzniak (2006) characterize three paradigms within the elementary geometry that is taught. These three paradigms are distinguished by the nature of the objects they study, the mode of validating statements, the organization of knowledge and the status accorded to drawing. The first two, called Geometry 1 (G1) and Geometry 2 (G2), model the geometry that can be taught in Grades 4–6. In G1, the objects are material objects: graphic or virtual traces on a screen. These representations of the tangible world are in fact a preliminary modeling of reality (Houdement 2007). The knowledge is organized in a pragmatic way for didactic or functional reasons. The validating of statements is done in relation to the tangible world, the representations being regarded as objects of the tangible world. Still according to Houdement and Kuzniak, the objects of G2 are ideal objects. The knowledge is organized according to a principle which is external to them, an axiomatic based on deductive reasoning. The validation of the statements is done in reference to this axiomatic through demonstrations. However, this geometry keeps a link with perception and reality, because the first founding axioms are compatible with the "perceived" (Houdement and Kuzniak 2006). In G2, the drawing has a double status: local modeling of the tangible space or representation of ideal objects.

This constitutes as a support for reasoning an aid to heuristics (Houdement 2007). This double status of drawing makes the work in G1 important for success in G2. The knowledge constructed in this framework in G1 has an immediate pragmatic function, allowing the pupils to solve problems in the tangible space, but it also prepares them for the heuristic approach required in G2. In this study, the proposed situations are placed in the framework of G1, the tangible space in which the problems are posed being of various sizes. This leads us to study the possible models in terms of the size of the space.

12.2.2. *The different spaces*

Depending on the space within which the subject is interacting, they develop different conceptual models. Berthelot and Salin (1992), Brousseau (2000), consider three types of space: micro-space, meso-space and macro-space. The micro-space is the space of interactions linked to the manipulation of small objects; the meso-space is the space of domestic movements. When a subject works in a micro-space, they are outside this space. They perceive the object they are studying as a whole. All relative positions between subject and object are possible, providing the subject with exhaustive information about this object. The activities are inexpensive and their effects are immediately perceptible (Brousseau and Galvez cited by Berthelot and Salin 1992). When the subject is confronted with tasks in the meso-space, they are part of the space in which they are working. They can move around quickly and then change their point of view. Their displacements confront them with different perspectives which can modify their perception of the spatial relations by which they identify objects in the micro-space. The work of Perrin-Glorian and Godin (2014, 2018) has emphasized the role of instruments in geometrical work, including whether or not a conventional measuring instrument, such as a graduated ruler, can be used. Like them, we hypothesize that approaching figures through magnitudes (lengths, angles), without measurements, facilitates entry into G2. This is why we associate with the work, in the meso-space, a choice of specific artifacts that do not allow measurement.

12.2.3. *Study on reasoning*

Bloch and Gibel (2011) propose a model of reasoning analysis based on TDS, which we rely upon for our analyses. The "milieu" structuring model, used in the development of this reasoning analysis model, is Bloch's (2006) model derived from Margolinas' (1995) model, but modified to account for the teacher's role in various levels of adidactic milieux.

In adidactic situations or situations with an adidactic dimension, the different stages of the situation can give rise to various types of reasoning, the function of which is linked to the level of the milieu: in terms of the reference situation, the confrontation with a heuristic milieu (objective milieu) allows for the elaboration of reasoning through action; the passage to a reference environment in the learning situation favors the elaboration of reasoning so as to justify the validity of the methods, and to establish the character of necessity for the properties that are used, more or less implicitly. For Bloch and Gibel (2011), identifying the functions of the reasonings helps the researcher identify the position of pupils in these levels of environments. We have shown (Gibel and Blanquart 2017) how, in the context of plane geometry, functions of reasoning are indeed related to the levels of settings, and how these functions also manifest these levels of settings. In doing so, we have retained three axes from the Bloch and Gibel (2011) model. The first axis of analysis focuses on the function of the reasoning produced, in relation to the level of the milieu to which they give evidence, in an action or formulation situation.

Concerning the action situations, to inform this first axis, we propose a classification of instrumented activities that can be carried out by the pupils in the tangible space, with the mediation of an artifact. To each category of action, we associate the function of the reasoning that may underlie it. This analysis a priori is theoretical. It distinguishes actions on the model from actions on the drawing, and then separates these actions according to whether they aim to obtain information about the object under consideration (epistemic mediation) or to modify this object (pragmatic mediation). Crossing these two criteria leads us to define four categories of actions: taking information about the model to be reproduced; taking information about the graphic object under construction (the drawing); modification of the model to be reproduced; and modification of the graphic object under construction. We then define the functions of reasoning that may be related to each of these action categories, based on the work of Gibel (2015). Finally, we group the identified functions of reasoning according to whether they relate to the use of artifacts, the characteristics of the model to be reproduced, the interpretation of feedback or the general organization of actions.

EXAMPLE.– The reasoning concerning the use of artifacts may have as its function either the choice of the artifact or the adaptation (or construction) of its usage patterns. Therefore, when a group of pupils uses the cardboard square to make an angle template, even though this technique is not part of the didactic repertoire of the class, we identify the production of a reasoning by construction of the schemas for the use of an artifact.

In the same way that we have proposed a classification of the functions of the reasonings specific to action, we have identified the reasonings that can characterize the formulation in the specific framework of figure reproduction. This work is not detailed here.

Concerning the second axis, in order to apprehend the possible reasonings produced by the pupils, we need to identify the objects they consider (on which they act or what they speak to), and the relations they perceive between these objects. In order to do this, we try to record the production and evolution of the different signs emitted by a pupil (or pupils), not independently of each other, but in terms of their relation to each other (Arzarello 2006), always in relation to the actions carried out (synchronic vision). In the abundance of signs produced, we retain the most salient ones that appear to us to be related to the geometrical activity in progress.

Finally, in connection with these observables, we seek to identify the knowledge mobilized by the pupils, the elements of the didactic repertoire of the class called upon and their evolution. This constitutes the third axis of study.

12.2.4. *The role of the teacher*

Using a bottom-up analysis, Bloch (1999) highlights the characteristics of the teacher's milieu and models it in adidactic levels in correspondence with the pupil's milieu. In his model, the teacher's milieu includes the "objective" elements of the milieu (pupils, situation, pupils' relationships and the different adidactic milieux), as well as the knowledge and skills that they bring into the situation.

12.2.5. *Problem statement*

With regard to these theoretical elements, our problem is as follows: in order to build the capacity for interventions and institutionalize knowledge, in plane geometry, how do Grade 4–6 teachers take into account the reasoning produced and the knowledge implemented by the pupils in the adidactic phases?

12.3. Methodology

12.3.1. *General principle*

Our research is based on the clinical analyses of interactions between pupils or between pupils and the teacher, in late elementary school and 6th grade classes, during an identical geometry sequence. This sequence, constructed by researchers, consists of two situations for the same knowledge, involving the rhombus, one in micro-space, and the "same" knowledge to be carried out in meso-space. The same descriptive sheet of the sequence was provided to the teachers for the experiments.

Fifteen sessions were observed in four classes. The body of raw data includes pupil productions and video recordings of all teacher interventions during the didactic phases, as well as recordings of actions and exchanges within some groups of pupils during the adidactic phases. Many of these data were transcribed in text form, and accompanied by photographs.

12.3.2. *The situations*

The situation in micro-space is a communication situation that takes place over two sessions. We do not go into detail here since our analyses focus mainly on the situation in meso-space.

The situation in the meso-space is a situation of reproduction of rhombuses cut out of a resistant and flexible card paper, whose sides measure between 65 cm and 80 cm. The session is structured in three stages: after a phase of devolution of the activity, the pupils are put in an action situation for about 20 minutes. Finally, they are grouped together to validate their productions and formulate their procedures.

For the action situation, the pupils are divided into groups in the classroom. Each group is given a model figure that cannot be moved outside a well-defined area. The instruction is to reproduce the figure as precisely as possible on the ground within a space reserved for tracing. This space is far from the area where the model is located. Next, in turn, each group explains to the whole class and to the teacher how they proceeded to trace the figure. The instruments and tools provided to the pupils are: string, scissors, a wooden bar about 2 m long, per group, as well as cardboard squares, felt-tip pens, different color chalks and blackboard dusters for erasing.

12.3.3. *Analysis methodology*

In order to bring elements that answer our problem statement, we analyze in greater detail the implementation of the situation within the meso-space, for a primary school class and a sixth-grade class. The pupils work in groups of three or four, and each group is studied separately. The study for a given group always follows the same methodology: we start by focusing our analysis on the pupils following the chronology of the session. We reconstruct the procedure used during the action phase, and then we identify the reasoning produced by the pupils within the group during the formulation phase. We then focus on the interventions of the teacher when they interact with this group: during the action phase (if it occurs) and then during the formulation phase, when the group in question presents its procedure to the rest of the class. We then obtain a working document from which we study the dynamics between the teacher's interventions, the evolution of their teaching project and the pupils' reasoning. For this purpose, we have summarized our line of questioning. The questions to which we wish to provide answers are interdependent – taken in isolation they are not very significant. Therefore, we divide the questions concerning the teacher's interventions into two parts. The first part includes questions concerning the ways in which the teacher "receives" the initial reasoning produced by the pupils. The second part groups the questions relating to the way in which the teacher makes these reasonings evolve. In the third part, we group together questions on the theme of learning memory.

EXAMPLE.– The questions in Part 3 are:

– Does the teacher make local institutionalizations? If so, how are they organized? Are links established between the different reasonings produced?

– Does the teacher synthesize or take stock of the new or old knowledge that was used during the session? In what form?

– Does the teacher use only oral communication or do they use transitional write-ups?

– Does the teacher produce or have the pupils produce a deferred write-up, once back in the classroom?

– Does the teacher make links with the existing didactic repertoire?

Moreover, observing teachers in action does not give access to the reasons for the choices made. This leads us to consider self-confrontation interviews with the teachers, in addition to the data collected during class time. The analysis of these interviews is based on the work of Bloch (1999) which we have presented. During a self-confrontation interview, the teacher is placed in a super-didactic level, in a reflective position with respect to the lower levels, adidactic and didactic. Through the video, they are led to question the elements of their milieu.

This allows us to analyze the teacher's relationship with the pupils in the situation. We organize this analysis according to the different levels: reference situation; learning situation; didactic situation; project situation. Finally, we compare this analysis resulting from the study of an interview with those resulting from the observation in class. We note the points of coherence or divergence that appear to us.

12.4. Conclusion

The analyses carried out show that in action situations, regardless of the size of the space, knowledge and reasoning are implemented by the pupils independently of the teacher.

We observed some pupils reinvesting this knowledge or reasoning in a new context, formulating it spontaneously in their exchanges with their classmates. For other pupils, the reasoning produced remained contextualized to a particular situation, and the knowledge underlying it was not reused, especially when the teacher did not help to reformulate or clarify further.

Therefore, on the teacher's side, different stages seem to participate in the dialectic between the production of reasoning by the pupils and the process of institutionalization:

– to identify the reasoning produced;

– to share knowledge and reasoning;

– to institutionalize locally "intermediate" knowledge;

– to decontextualize, generalize;

– to create links between the reasoning and knowledge that "circulate" and old knowledge.

The recognition of the reasoning produced by the pupils during the action phase requires active observation of the pupils by the teacher. The 4th/5th grade teacher seems to us to be more successful in this observation than the one in 6th grade: we hypothesize that the prior analysis of the situation that they have carried out, coupled with a very precise devolution, facilitates this observation and helps them to conduct the formulation phase. The sharing of knowledge and reasoning aims at the appropriation by all the pupils of the reasoning produced by some of them, and of the knowledge that has circulated during the activity phase. We observed that in the meso-space, sharing is favored by the actions of the two teachers observed, who value, compare and put into perspective the different reasonings produced by the pupils. In this way, they organize a real institutionalization process. The two teachers also institutionalize intermediate knowledge that will participate in the elaboration of the knowledge text. However, decontextualization and generalization are initiated in both classes without always being very explicit. Finally, the 4th/5th grade teacher establishes relationships between knowledge that is present or missing at different points in the sequence. We observe fewer of these links in the class of the sixth-grade teacher who, during an interview, points out the difficulty of relying on a common didactic repertoire in a class where the pupils come from several schools.

Therefore, many of the teachers' choices can be questioned in relation to the constraints of their institution. This is work that we have begun, and which could be pursued in order to access the teacher's activity in terms of over-didactic levels.

12.5. References

Arzarello, F. (2006). Semiosis as multimodal process. In *Sémiotique, culture et pensée mathématique*, Radford, L. and D'Amore, B. (eds). Revisita Latinoamericana de Investigación en Matemática Educativa, Colonia Santa María la Ribera.

Berthelot, R. and Salin, M.H. (1992). L'enseignement de l'espace et de la géométrie dans la scolarité obligatoire. PhD Thesis, Université Sciences et Technologies-Bordeaux I.

Blanquart, S. (2020). Raisonnements géométriques d'élèves de CM, duos de situations, rôle de l'enseignant. PhD Thesis, Université de Paris [Online]. Available at: https://tel.archives-ouvertes.fr/tel-03242768/ [Accessed 20 January 2022].

Bloch, I. (1999). L'articulation du travail mathématique du professeur et de l'élève dans l'enseignement de l'analyse en Première scientifique. *Recherches en didactique des mathématiques*, 19(2), 135–193.

Bloch, I. (2006). Quelques apports de la théorie des situations à la didactique des mathématiques dans l'enseignement secondaire et supérieur. HDR synthesis note, Université Paris 7.

Bloch, I. and Gibel, P. (2011). Un modèle d'analyse des raisonnements dans les situations didactiques : étude des niveaux de preuves dans une situation d'enseignement de la notion de limite. *Recherches en didactique des mathématiques*, 31(2), 191–228.

Brousseau, G. (1998). *Théorie des situations didactiques*. La Pensée Sauvage, Grenoble.

Brousseau, G. (2000). Espace et géométrie. *Actes du séminaire de didactique des mathématiques*. Département des Sciences de l'éducation de l'Université de Crète à Réthymnon, 67–83.

Gibel, P. (2015). Mise en œuvre d'un modèle d'analyse des raisonnements en classe de mathématiques à l'école primaire. *Éducation et didactique*, 9(2), 51–72.

Gibel, P. and Blanquart, S. (2017). Favoriser l'appropriation des propriétés géométriques des quadrilatères à l'école primaire : étude d'une situation d'apprentissage dans le méso-espace. *Revue des sciences de l'éducation*, 43(1), 37–84.

Houdement, C. (2007). À la recherche d'une cohérence entre géométrie de l'école et géométrie du collège. *Repère IREM*, 67, 70–84.

Houdement, C. and Kuzniak, A. (2006). Paradigmes géométriques et enseignement de la géométrie. *Annales de didactique et de sciences cognitives*, 11, 175–193.

Margolinas, C. (1995). La structuration du milieu et ses apports dans l'analyse a posteriori des situations. In *Les débats de didactique des mathématiques*, Margolinas, C. (ed.). La Pensée Sauvage, Grenoble.

Margolinas, C. (2004). Points de vue de l'élève et du professeur. Essai de développement de la Théorie des Situations Didactiques. HDR synthesis note, Université de Provence, Aix-Marseille I.

Perrin-Glorian, M.J. and Godin, M. (2014). De la reproduction de figures géométriques avec des instruments vers leur caractérisation par des énoncés. *Math-école*, 222, 23–36.

Perrin-Glorian, M.J. and Godin, M. (2018). Géométrie plane : pour une approche cohérente du début de l'école à la fin du collège [Online]. Available at: https://hal.archives-ouvertes.fr/hal-01660837v2 [Accessed 29 November 2021].

Schneider, M. and Mercier, A. (2008). Situation adidactique, situation didactique, situation-problème : circulation de concepts entre théorie didactique et idéologies pour l'enseignement. *Actes du colloque didactiques : quelles références épistémologiques*, AFIRSE Bordeaux, 1–14 [Online]. Available at: https://hal.archives-ouvertes.fr/hal-01995384 [Accessed 29 November 2021].

13

When the Teacher Uses Common Language Instead of Geometry Lexicon

13.1. Introduction

The understanding and proper use of mathematical lexicon, particularly that of geometry, are part of the expectations of the programs for each learning cycle. The official instructions insist on the acquisition of the lexicon by the pupils no matter the discipline (Eduscol 2015) and a fortiori in mathematics (MEN 2016). The idea is to promote the reading, the exploitation and the communication of results from various representations, to help learning, to argue, to describe and to better conceive the objects studied.

The instructions for the French learning cycle equivalent for Grades 1–3 state that "in geometry as elsewhere, it is particularly important that teachers use precise and appropriate language, and introduce the appropriate vocabulary during manipulations and action situations where it makes sense to pupils, and that pupils are gradually encouraged to use it" (MEN 2018, p. 72, *translated by author*). Therefore, these prescriptions encourage teachers to be terminologically accurate in terms of geometry in order to familiarize their pupils with these uses, but does this mean that they always need to implement this recommendation?

During the analysis of six sessions proposed to allophone secondary school pupils in *Unité pédagogique pour élèves allophones arrivants* (UPE2A), a pedagogical unit for allophone pupils (Hache et al. 2020), we have identified in the teacher's discourse the usage of common language terms to designate geometrical concepts. Following this study, we classify these occurrences into five categories,

---

Chapter written by Karine MILLON-FAURÉ, Catherine MENDONÇA DIAS, Céline BEAUGRAND and Christophe HACHE.

based on a verbal interaction situation, and we seek to verify whether we find these same phenomena in three geometry lessons taught in a regular sixth-grade class (Beaugrand 2019).

13.2. An attempt to categorize the uses of common vernacular terms in place of geometry lexicon terms within teacher discourse

13.2.1. *The phenomenon of didactic reticence*

First of all, we could see that at certain moments the UPE2A teacher refused to use the appropriate term, in order to make sure that her pupils knew it, as we can see in this extract:

> Teacher: When we make a circle piece, do you know how to say it?

Here, the teacher wants to lead her pupils to find the expression "the arc of a circle", which leads her to use a periphrasis in her question: "piece of a circle", so as not to reveal the answer. Similarly, in the excerpt below, the teacher insists on using the term "line" instead of the phrase "straight line":

> Teacher: What kind of line is the mediator? What type of line is it, the mediator? What is this line, what type of line?

In these examples, the teacher refrains from using certain knowledge (the precise names of the concepts mentioned). We might think that she wanted to give the pupils the possibility of proposing them themselves, and thus facilitate their memorization. These are therefore phenomena of "didactic reticence":

> As Brousseau (1998) showed very early on, the teacher cannot "tell knowledge" to the pupil directly, even in the most classic and directly transmissive forms of teaching, because then there would be no guarantee of the pupil's effective understanding and real use of this knowledge. The teacher is thus led to "withhold information", to be *reticent* (Sensevy and Quilio 2002; Sensevy 2007). We employ the word *reticent* here in the sense of the old French, of a "voluntary omission of what could or should be said", still active in English (Sensevy 2008, p. 45, *translated by author*).

We also find several phenomena of this type in the lessons observed in the ordinary classroom setting:

Teacher: Remember the difference between a circle and something else?
Pupil: A disk
[...]
Teacher: We saw it in the lesson, right? What a piece of a circle was. It's called a...
Another pupil: An arc
Teacher: An arc of a circle.

13.2.2. *The phenomenon of semantic analogy: comparison with common concepts to construct meaning for mathematical knowledge*

The UPE2A teacher sometimes uses concepts from everyday life, and therefore, a priori known to the pupils, to construct the meaning of geometric knowledge.

Some of these connections are based solely on the common (or at least close) form of these two concepts, as in this extract where the teacher uses a comparison proposed by a pupil:

Teacher: An arc (P draws an arc)
Pupil: Like the moon?
Teacher: Yes, like the moon.

Sometimes the analogies used by the UPE2A teacher are also based on a semantic derivation. As such, in this extract, the teacher tries to facilitate the memorization of the past participle "aligned" by linking it to the noun, a priori better known to the pupils (although polysemous, etc.), "line":

Pupil: Aligned
Teacher: Aligned, what does aligned mean?
Pupil: Like this
Teacher: Yes, they are all on the same line. The "lines" are "aligned".
On a straight line.

We can note that here the connection could mislead the pupils since the term "line" – "continuous line, whose extent is practically reduced to the sole dimension of length" according to the Larousse dictionary – differs somewhat from the concept of straight line: on the one hand, because a line has a certain thickness (even if it is very weak), but moreover because it is not necessarily a "straight line".

It should be noted that in the expression "bus line" (probably well known to the pupils), the line in question is generally curved, etc. In the ordinary classroom, we also find in the teacher's discourse support for the memorization for geometric lexicon terms found in everyday language:

> Teacher: [about "arc of circle"] And why an arc? Because it looks like
> Pupil 1: Like an arc
> Teacher: Like an arc. And a bow, when it's taut, it's made taut by what?
> Pupil 2: A bowstring
> Pupil 3: String
> Teacher: We also call a bowstring a "cord". That's why it's called that[1].

This comparison with the bow used for hunting can effectively allow pupils to memorize the expression "arc of a circle" and the term "chord". However, a few minutes after this exchange, we can note the confusion of a pupil who will make a semantic shift and a lexical transposition, in a mathematical context, of "string", a term which in everyday life designates a thin rope:

> Teacher: Not a string, a cord. Yeah, you don't want to mix it up, do you? Don't tell me there are strings in a circle.

13.2.3. *The phenomenon of lexical competition: use of common vernacular terms to designate common concepts*

In the UPE2A body, the teacher sometimes uses terms from common vernacular in a context where pupils might think that she is talking about mathematical objects, when this is not the case.

This phenomenon appears in particular during geometry lessons where the teacher sometimes refers to the underlying material activity, and in particular to the tracing of figures:

> Teacher: We take our ruler, and we take our pencil to make a line.

Here, it is not the line segment itself that the teacher is trying to designate, but rather its representation on the sheet of paper, which can effectively be designated by the term "line". We can note that the word "line" is used as a "line, an elongated mark drawn on paper or on any surface" (Larousse, *translated by author*), and as such, can be used to designate different object representations in mathematics, as in

---

[1] A bow can be translated in French by *un arc*, like an arc of a circle (*un arc de cercle*). A bowstring can also be called *une corde* and a chord in a circle is also called *une corde*.

the following intervention by the teacher where the word "line" is no longer associated with a line segment, but with the arc of a circle: "I want a compass line."

In the ordinary classroom setting, we also find the use of terms from everyday language in connection with the tracing of figures, as in this excerpt where the fact of designating the endpoint of the line segment by the term "end" makes it possible to insist on the idea of terminating the tracing:

> Teacher: The ruler. And then my pencil, I put it down and I let it press until I get to the end of the line segment.

A similar case occurs in the following example, where the teacher uses "piece of a circle" (which here means a piece of the circle line) when the expression "arc of a circle" was recalled earlier:

> Teacher: I don't have to draw the whole circle. I just have to draw a piece of the circle.

Other interventions by the teacher can fall into this same category, such as this one, taken from one of the UPE2A sessions: "I want the line segment in the middle of the sheet." Since the sheet is rectangular, mathematically the "middle of the sheet" does not really make sense: the term "center" seems more appropriate to designate the point at equal distance from the four vertices of the rectangle. However, here, the term "middle" should not be taken in the mathematical sense but in the usual sense – "A place equally distant from all the points around or at the ends of something" (Larousse, *translated by author*) – and it can therefore be applied to objects with a dimension greater than 1: "the middle of the table", etc. However, we may wonder what reception pupils have of this type of usage by the mathematics teacher, especially in a lesson on the circle where they will repeatedly revoke pupils who speak of the "middle of the circle" by explaining that the term "middle" is reserved for segments. It should be noted that the exact same example is found in one of the sessions in the ordinary classroom setting, when the teacher indicates where to place one of the frames: "Well, yes, in the middle of the page."

13.2.4. *The phenomena of repeating pupil formulations*

Sometimes, the teacher has to repeat the pupils' less than rigorous or even erroneous expressions, in particular, in order to correct them or to ask for a reformulation. Let us quote this example taken from the UPE2A sessions:

> Pupil: Straight triangle
> Teacher: We don't say straight triangle, but a right-angled triangle.

By repeating a pupil's incorrect statement, the teacher makes sure that everyone has heard it and emphasizes the difference with the expected expression. In the regular classroom, the teacher will also repeat the pupils' erroneous statements several times to make they are aware of their errors. Therefore, on five occasions, they pretended to be indignant about the use of the word "round":

> Pupil: A round
> Teacher: A what? A what? What did I hear? Who said that horrible word to me? How horrible! How horrible! A round, that doesn't make any sense!

In the same way, at another moment, it will be enough for the teacher to repeat the erroneous proposal of a pupil so that the latter realizes their error and rectifies the expression used by themself:

> Pupil: Um, at the intersection of... the first intersection of... Rayan and Lucy, below (Pupil is talking about one of the points of intersection between the circle at the center of Rayan's house, and the circle at the center of Lucy's house)
> Teacher: The intersection of Rayan and Lucy?
> Pupil: No. The intersection.

There is also another situation in which the teacher is led to take back an erroneous formulation proposed by a pupil: when they are afraid of interrupting the class's reflection by leading the pupils to focus on the form of their speech. In this case, the teacher will generally prefer to postpone the introduction of the correct terminology to another moment in the sequence, for example, during the institutionalization phase (Brousseau 1998).

13.2.5. *The phenomenon of didactic repression*

Finally, teachers sometimes refrain from using certain terms in the mathematical lexicon for fear that their pupils will not understand them and that this will disrupt their mathematical activity.

This is the phenomena of "didactic repression":

> The teacher forbids himself the introduction into the classroom of a certain concept and/or the associated lexicon for fear of losing his pupils. We will call this censorship that the teacher imposes on himself in order to facilitate his pupils' learning, didactic "repression" (Millon-Fauré 2017, p. 170, *translated by author*).

This phenomenon is to be distinguished from didactic reticence, insofar as the purpose of this self-censorship on the part of the teacher is radically different in these two cases: in the case of the phenomena of didactic reticence, the aim is to give the pupil the opportunity to evoke on their own the objects of knowledge that have been withheld, in order to encourage their appropriation, whereas in the case of the phenomena of didactic repression, the teacher prefers to withhold these terminologies, which are judged to be excessively complex, from being introduced into the environment.

We did not find any phenomena of didactic repression in the six geometry lessons observed in UPE2A. However, we did find one case in a lesson on calculations in the same class. In order to simplify a fraction, the teacher asked: "I want you to divide up and down." By using the expressions "up" and "down", the teacher avoided the use of terms specific to mathematics: "numerator" and "denominator", which are certainly less familiar to the pupils. Similarly, in the ordinary classroom setting, we noted a case that could be related to a phenomenon of didactic repression:

> Teacher: Is this a whole disc?
> Pupil: No
> Teacher: Well, no, because it's missing this whole part here

We are talking about an angular sector, but this term is not known to the pupils and is not included in the program for this learning cycle. The teacher therefore prefers to simply say that it is a disc from which a part has been removed, so as not to complicate their discourse.

However, it had been shown previously that the phenomena of didactic repression could mislead the pupils. In fact, given that the terms used (generally coming from everyday language) do not cover exactly the same reality as the terms (in our case, from the mathematical lexicon) that they replace, they do not correspond exactly to the designated concept. For example, during a geometry lesson offered in a UPE2A (Millon-Fauré 2017), the teacher had designated their solids by the term "objects", in order to simplify their discourse. The pupils had not grasped then the distinction between solids and the plane figures corresponding to their faces. This then led them to confuse "cube" with "square", or "cuboid" with "rectangle".

13.3. Conclusion

This study shows us that, in both the regular class and the UPE2A, the teacher uses, at times, terms from the common vernacular in place of terms from the expected geometry lexicon and that, in both contexts, these uses seem to respond to the same five main types of motivations: phenomena of didactic reticence reliance on everyday language to construct mathematical concepts or to facilitate their assimilation, reference to everyday concepts (in particular traces of material activity) but which can lead to confusion with mathematical concepts, repetition of erroneous formulations by the pupils and the phenomenon of didactic repression. The teacher resorts to these professional gestures to facilitate communication with their pupils and to facilitate their learning, and it would be interesting to study the frequency of their appearance according to the teacher's didactic intentions. However, in addition to the difficulties previously mentioned concerning the misunderstandings that these uses could lead to, we wonder how, under these conditions, the pupils manage to identify the moments when they must use the geometry lexicon and not common vernacular terms. Do they really manage to distinguish between these two types of lexica, if the teacher employs one and then the other alternately? How can we help them to perceive the specificities of the mathematical lexicon and what work could we put in place to facilitate this learning?

13.4. References

Beaugrand, C. (2019). Transposition des démarches du français sur objectifs spécifiques en contexte scolaire. PhD Thesis, Université de Paris 3-Sorbonne, Paris.

Brousseau, G. (1998). Glossaire de quelques concepts de la théorie des situations didactiques en mathématiques [Online]. Available at: http://guy-brousseau.com/wp-content/uploads/2010/09/Glossaire_V5.pdf [Accessed 28 January 2022].

Eduscol (2015). La maîtrise de la langue française : un plan d'action global [Online]. Available at: https://www.education.gouv.fr/la-maitrise-de-la-langue-francaise-un-plan-d-action-global-10478 [Accessed 28 January 2022].

Hache, C., Mendonça Dias, C., Millon-Fauré, K., Azaoui, B. (2020). Everyday terms in mathematical classroom interactions: Case study with multilingual immigrant learners. In *ETC 7 – Seventh ERME Topic Conference: Language in the Mathematics Classroom*, 18–21 February 2020, Montpellier University.

MEN (2016). Mathématiques et maitrise de la langue. *Eduscol, ressources transversales* [Online]. Available at: https://cache.media.eduscol.education.fr/file/Ressources_transversales/99/6/RA16_C3C4_MATH_math_maitr_lang_N.D_600996.pdf [Accessed 26 January 2022].

MEN (2018). Programme du cycle 2. *Journal officiel du 12 mars 2015*.

Millon-Fauré, K. (2017). *L'enseignement des mathématiques aux élèves allophones*. La plaine Saint Denis, Connaissances et savoirs, Paris.

Sensevy, G. (2007). Des catégories pour décrire et comprendre l'action didactique. In *Agir ensemble : l'action didactique conjointe du professeur et des élèves*, Sensevy, G. and Mercier, A. (eds). PUR, Rennes.

Sensevy, G. (2008). Le travail du professeur pour la théorie de l'action conjointe en didactique. *Recherche et formation*, 57, 39–50.

Sensevy, G. and Quilio, S. (2002). Les discours du professeur. Vers une pragmatique didactique. *Revue française de pédagogie*, 141, 47–56.

14

The Development of Spatial Knowledge at School and in Teacher Training: A Case Study on *1, 2, 3... imagine!*

14.1. Introduction and research question

In this chapter, we present our reflections on teaching and learning about space. Our contribution to this textbook, *1, 2, 3... imagine!,* is focused more on the side of spatial knowledge.

As the Ontario Ministry of Education states:

> Most of us have been taught to think and talk about the world using words, lists, and statistics. These are useful tools but they do not come close to telling the full story. Thinking spatially opens the eye and mind to new connections, new questions, and new answers (Center for Spatial Studies, 2014, p. 5).

The spatial knowledge discussed in this excerpt is part of the curriculum not only in Ontario, but also in Quebec (Government of Quebec 2006) in the section entitled "Geometry: Geometric Figures and Spatial Sense", in France (MENIS 2020) and in many other countries around the world. Overall, we will call spatial knowledge (SK) "the knowledge [...] that allows one to master the anticipation of the effects of one's actions on space, their control, and the communication of spatial information" (Berthelot and Salin 1992, p. 9, *translated by author*). In mathematics, this

---

Chapter written by Patricia MARCHAND and Caroline BISSON.

knowledge is concomitant with geometric knowledge (GK). These two types of knowledge are both inseparable and distinct.

Although SK is part of the primary mathematics curriculum in several countries, it is not necessarily integrated into the curriculum in a consistent manner, or as thoroughly, from one school reform to the next. Overall, it is possible to argue that they have been neglected in the educational programs of various countries (Moss et al. 2016; Verdine et al. 2017; Perrin-Glorian and Godin 2018). However, as Salin already advocated in 2013, previous authors mention that there is currently a shift towards a greater consideration of SK in mathematics education. The goal of this surge is not only to substantially include SK in the curriculum but also to make it a pillar in itself. This growing concern with SK is based on research findings highlighting its correlation with pupil success rates in geometry (Kalogirou and Gagatsis 2012; Battista 2008; Gaulin 1985), problem solving (Hegarty and Kozhevnikov 1999; Clements and Sarama 2011; van Garderen 2006; National Council of Teachers of Mathematics (NCTM) 2006), mental arithmetic (Kyttälä and Lehto 2008), algebra (Tolar et al. 2009), Mathematics in general (NCTM 2005; Cheng and Mix 2012; Shumway 2013; Verdine et al. 2017; Sorby and Panther 2020), beyond Mathematics and throughout Science (Wai et al. 2009; Chastenay 2015; Davis et al. 2015). However, in previous research, it was possible for us to observe that, for the same classroom activity, the didactic choices made upstream of the class session strongly colored the development of SK and the success rate of pupils (Marchand 2006a, 2006b, 2009, 2020; Munier et al. 2011; Marchand and Braconne-Michoux 2014; Marchand and Munier 2021). This observation led us to question the conditions that should be put in place to enhance this development. Moreover, it is known that pupils face several difficulties with regard to these SKs, both at the primary and secondary school levels (Parzysz 1988; Berthelot and Salin 2000; Clements 1999; Salin 2008; Mithalal 2014).

Moss et al. (2016) thus identify five reasons why it seems relevant to focus on SKs: 1) they are intimately related to the development of mathematical thinking; 2) they are "malleable", in the sense that they can develop in any pupil, regardless of age, even as an adult (Uttal et al. 2013); 3) they are an important predictor of pupils' future success in science, technology, engineering and mathematics (STEM); 4) they are currently a blind spot in our educational programs; and 5) they can be viewed from a variety of perspectives, thus making mathematics accessible to all (an equity issue). And more broadly, the 2002 report from the Commission on Mathematics Education (led by Kahane) identified four reasons for teaching geometry, one of which was to develop a vision of space and its representations.

The correlation between SK and mathematical thinking is well established, but like Mulligan et al. (2020), our interest is in furthering this reflection on ways to develop them in the classroom, which, in turn, are still under-explored (Marchand

and Sinclair 2017). In this chapter, we focus on a possible activity that can be used to develop SK. An analysis of this activity will be conducted, and we will outline four teaching sequences derived from this activity. The first three were designed for primary school pupils in Quebec, more specifically for the 1st (6–8 years old), 2nd (8–10 years old) and 3rd cycles (10–12 years old) of primary school, and the fourth was designed for pupils in the bachelor's degree program in special education at our university institution. The conceptual framework is first presented, and then the activity of Yackel and Wheatley (1990) is exposed and analyzed in relation to it, as well as the didactic variables that can be used. This is followed by a description of the various sequences developed in the framework of our work and the results of the two experiments (primary and teacher training).

14.2. Conceptual framework

The concept of SK, like spatial ability or spatial thinking, is difficult to pin down since it is an "umbrella concept" referring to a constellation of knowledge across a wide variety of applied domains (mathematics, science, sports, arts, etc.), as mentioned by Uttal et al. (2013) and Verdine et al. (2017). SK groups knowledge that physically (concrete actions) or mentally (mental images) leads the learner to control, anticipate, and communicate states, displacements, or deformations of objects related to tangible or geometric space, in two or three dimensions (Berthelot and Salin 2000; Battista 2007; Marchand 2020).

The "spatial orientation" component refers to orientation situations in space that require pupils to be able to find their way around the space, locate an object and navigate through it. This facet also requires the use of a frame of reference (laterality, coordinates of a plane, cardinal points, etc.) as well as a recourse to extrinsic spatial relationships between the different objects that make up this space (Clements 1999; Panaoura et al. 2007; Uttal et al. 2013).

The "spatial organization" component refers to situations of organization in space that involve, on the part of pupils, the production and study of Mental Images (MI) of tangible or geometric objects. It involves a fine analysis of the articulation of these, that is, the intrinsic spatial relations (parts-parts/parts-object relationship) (Clements 1999; Panaoura et al. 2007; Uttal et al. 2013).

For the same activity, the "orientation" and "organization" components can be solicited. However, it must be assumed that the use of one does not necessarily lead to the development of the other. A pupil may be able to find his or her way around the school or his or her neighborhood, but may not be able to imagine a triangle mentally rotated to form a solid of revolution (Braconne-Michoux 2013). Let us also mention that the development of SK is not limited to the school experience but goes beyond the walls of the school, through other experiences involving their relations to tangible space, such as sports, music or even video games (Berthelot and Salin 2000). Moreover, their development extends into adulthood and is conceivable in the school setting (Uttal et al. 2013; Davis et al. 2015).

With these observations in mind, we describe below the didactic theoretical frame supporting this project (Marchand 2020). We thus present three foundations of our framework: the components that define SK; the levels of abstraction linked to their development in school and the essential variables of the activity for their valorization. These guidelines[1] are a construct called "activity generating structure" (AGS), which does not claim to be exhaustive, but which aims to deepen the theoretical reflection surrounding the development of SK in schools (Figure 14.1).

SK
- Locating
- Referential
- MI
- Articulation

Level of abstraction
- Archaeological
- Photographic
- Scenographic

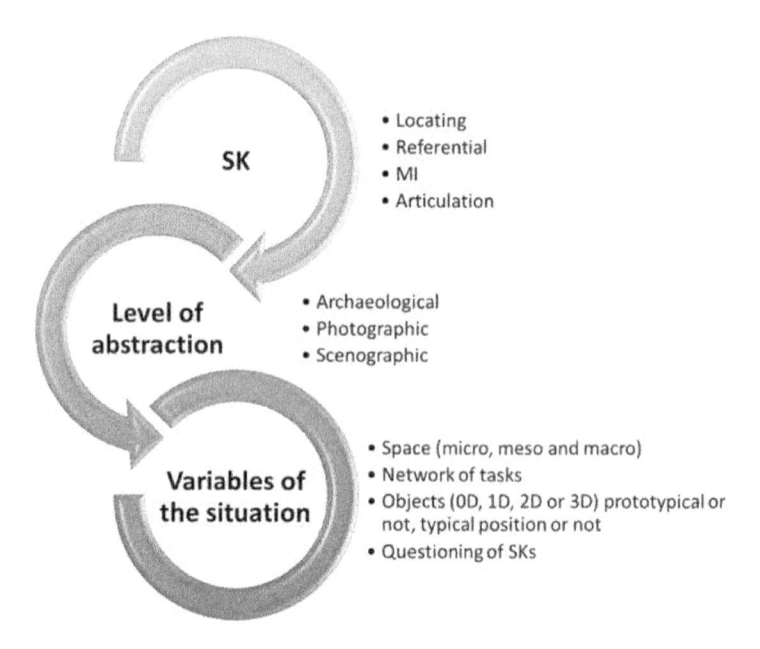

Variables of the situation
- Space (micro, meso and macro)
- Network of tasks
- Objects (0D, 1D, 2D or 3D) prototypical or not, typical position or not
- Questioning of SKs

Figure 14.1. *Diagram of the activity generating structure (AGS)*

14.2.1. *Components set to address SK in primary school*

The components that we have set to deal with SK have been identified previously, but we recall them here by situating them within our theoretical tool. Therefore, SK can be broken down into two components: orientation and

---

[1] The term "guideline" is used here analogously. The guidelines represent a set of didactic reference points allowing us to delimit, to circumscribe, the SK in mathematics class.

organization. These two components involve, respectively, extrinsic and intrinsic spatial relationships (Uttal et al. 2013). Each of these components has two predominant subcomponents: the "orientation" component involves the localization using a frame of reference, and the "organization" component deals with mental images (MI) and their articulation. These two components are not mutually exclusive; they have their own specificities and each contributes to the development of SK. On the other hand, the "spatial organization" component, involving MIs and intrinsic relationships, represents a blind spot in several educational programs, including the one in Quebec (Gouvernement du Québec 2006).

Yet work in psychology has demonstrated, and has done so for many years, a correlation between the development of SK related to mental object rotations (MI and articulation), and pupil performance in primary and secondary school (Hawes et al. 2015). This initial theoretical foundation is not sufficient in itself, to create and analyze classroom activities that solicit and develop SK. We must also qualify these two components according to whether they feature tangible or geometric objects, static or dynamic, and whether they require pupils to do tangible or mental work. These characteristics are addressed with the second foundation according to Battista's (2007) theory of abstraction.

14.2.2. *Levels of abstraction that value SK*

The development of SK and GK represents an iterative process between action (physical and mental) on the one hand, and reflection and abstraction on the other (ibid.). A first level of abstraction that we call "archaeological" allows pupils to become aware of the objects around them. Battista compares this first level to the work of an archaeologist who analyzes fragments by defining their characteristics and linking them to those already identified in order to categorize them (ibid.). By analogy, the pupils' lived experiences, using manipulative tasks, allow them to isolate certain perceptual properties of objects in order to be able to recognize and categorize them. The next level of abstraction, which we entitled "photographic", is called upon when the objects have been studied sufficiently by the pupils to enable them to mentally represent them in their absence (ibid.). Therefore, here, it is a question of thinking based on perceptual recognition which, consequently, remains linked to the objects observed. The MIs involved here are static and reproductive (Piaget and Inhelder 1977), and they provide access to a single perspective (the observed one). At this level, following the parallel development of GKs, images become concept-images conveying their geometric, spatial and metric properties (Battista 2007). The third level of abstraction, described as "scenographic", is characterized by the analysis of the MI of the "photographed" objects. It requires, on the part of the pupils, to "take a step back" in relation to the MI in order to identify their structure and composition. They must, like a scriptwriter, imagine the characters (here the objects), their characteristics (geometric, metric and intrinsic properties), their interactions (extrinsic), as they move through space (ibid.). The images are concept-images, which are now kinetic, transformative and even anticipatory (Piaget and Inhelder 1977). Pupils at this level are able to move them around mentally, able to consider different perspectives, to compose and transform them, just as we would do physically. This second foundation allows us to situate SK according to whether it is physical (static or kinetic) at the archaeological level, mental/static at the photographic level, or mental/kinetic at the scenographic level. In this sense, Marchand (2020) has empirically found that moving to the third level appears to be conducive to the development of SK. A third and final foundation, in terms of task-related variables, was considered in the development of this AGS.

14.2.3. *Main variables in situations where SK is valued*

Four variables seem unavoidable to us since they are recurrent in our work and that of other researchers, as we shall see.

The first variable is the conception of space (micro, meso and macro) (Brousseau 1983; Berthelot and Salin 1992). Briefly, micro space represents the space of small objects that can be manipulated, or the space of the sheet of paper. This is the space usually advocated in mathematics classrooms. The meso space is a space where the movements of the pupil or objects are considered, and which is entirely accessible through the field of view. Moving between micro and meso space seems a promising avenue for conceptualizing both SK (Brousseau 1983; Braconne-Michoux and Marchand 2021) and GK. Macro space is a space too large for the pupil "to take it all in with a glance" (Brousseau 2001, p. 6, *translated by author*). Only a local and partial view is possible. Berthelot and Salin (2000) provide an example where pupils are asked to produce a map of a road network. The pupils must thus use maps of local spaces by exploiting the MIs of macro-space and their material representation of micro-space.

The second variable is the nature and network of tasks that can be offered to pupils. A list of tasks was generated for this project based on the work of Berthelot and Salin (2000), Davis et al. (2015) and Moss et al. (2016): observing or touching, identifying, describing; situating, locating, mapping, coordinating perspectives; moving, assembling, decomposing, folding, rearranging; transforming, deforming, sectioning, scaling; constructing, depicting; anticipating, searching. This grouping of tasks remains exploratory at present, but we wanted to make a first sorting given the panoply of tasks listed[2]. One of the main results obtained in our previous work is that the network of tasks is a determining factor in the level of difficulty of the

---

2 Davis et al. (2015) alone identified 29 tasks.

activity. For example, the following two timelines seem promising for developing SK: anticipating the material needed before any construction, building and then describing the resulting solid (Marchand 2006b, 2009); observing a picture composed of Tangram pieces in a limited time and reproducing it in its absence (Yackel and Wheatley 1990). In these two cases, the anticipation task is solicited differently, but it is always shifted (before or after) the manipulation or resolution (Braconne-Michoux and Marchand 2021). Moreover, we find this constitutes a gap of the activities set up by other authors in this textbook. Among the activities proposed in the other chapters, we think to Celi's activity, in which pupils have to manipulate and describe geometric figures blindly, or Douaire et al's activity, in which pupils have to find the solid after observing it first, and then touching it afterwards. In these two cases, there is a shift in the tasks asked of the pupils, which seems to provoke the passage from one level of abstraction to another.

The third variable is the nature and size of the objects involved. The objects can be 0D (the point), 1D (line), 2D (figure) or 3D (object), and they can be positioned in a typical or non-typical way and be prototypical or not (isosceles trapezoid versus any). This variable plays an important role in geometric conceptualization. We hypothesize that it also plays a significant role in the development of SKs since it refers to the appearance of objects, their positioning in space and their articulation. By varying the nature and dimension of tangible or geometric objects, the creation and comparison of MI seem to be enhanced. Finally, the fourth and last variable targeted by this theoretical tool is the topic of institutionalization questioning. It should focus on the SK solicited by the activity. This questioning allows pupils to become aware of their own SK, to share with others, to hear other reasoning, thereby enriching their own SK.

These variables represent those that we have considered, up to now, as essential for developing SK; however, it is possible to add other variables so as to generate activities aimed at developing SK according to the desired framework and depth of analysis. The following section proposes an activity valuing SK development and its analysis according to this conceptual framework.

14.3. Presentation of the activity 1, 2, 3 ... imagine!

When we looked at the development of SK, it was possible to find examples of activities that the author mentioned were conducive to the development of SK. However, there were few theoretical guidelines for understanding why.

For example, Yackel and Wheatley (1990) propose a situation using a Tangram, in which pupils must observe a picture made up of Tangram pieces for three seconds and then reproduce it in its absence. This activity does seem to develop SK, but why is it more so than another? The AGS theoretical tool is one way of answering this question, but let us first turn our attention to the activity proposed by Yackel and Wheatley (1990).

Pupils are asked to observe a generated image of Tangram pieces projected onto a screen at the front of class for 3 seconds, with their hands behind their backs so that they do not manipulate the pieces during observation. The image is then hidden and they must reproduce it with their Tangram pieces. The image can be observed in this way two to three times, and the teacher's intention is clearly mentioned to the pupils at the beginning:

> I would like to know how you manage to reproduce the image on your desk. How do you manage to remember the image in your head? And how do you build it on your desk? In what order? How were the figures positioned in space and in relation to each other?

A session can be composed of three images like those presented, in a reduced model, in Figure 14.2, and these questions are repeated after each image.

Figure 14.2. *Examples of images that can be processed in this activity. For a color version of this figure, see www.iste.co.uk/guille/tangible.zip*

Table 14.1 shows the top techniques (τ) for each of the tasks (T) that this activity solicits according to our a priori analyses and our empirical data (Marchand 2020).

The order of presentation should not be interpreted as a chronology or hierarchy; it is a list. The same pupil can use a combination of these techniques to solve each of these tasks. The purpose of this list is to make the possible techniques[3] explicit. These techniques illustrate the complexity of this activity wherein different components and spatial relationships must be considered simultaneously in order to successfully complete the task.

---

[3] Having experimented with this activity in primary school and in teacher training in Quebec, we can say that these techniques are used by everyone. Therefore, they seem to be essential in order to carry out this activity.

Tasks (T)	Techniques (τ)
T₁: Successive observations of the Tangram	τ1: Focus on one or two pieces in the image, remembering their positioning and articulation (spatial relationships). τ 2: Identify the number and choice of pieces. τ 3: "Take" a picture of the image. τ 4: Scan the image horizontally or vertically. τ 5: Create a path or story for yourself. τ 6: Focus first on the articulation existing between the pieces, and in a second step, on the position of these pieces or vice versa.
T₂: Reproduction and adjustment of the Tangram	τ7: Mentally find the look and composition of the Tangram. τ 8: Turn, position, and flip the pieces to match the mental image created and the concrete reproduction of the Tangram. τ 9: Identify the pieces needed for reproduction. τ 10: Compare and adjust our production according to the chosen mental image.
T₃: Explanation of the technique used	τ11: Recall and verbalize the technique used. τ 12: Use appropriate spatial and geometric vocabulary. τ 13: Listen to others' techniques and identify similarities and distinctions with our own. τ 14: Understand and appropriate the techniques of others when revisiting future cases.

Table 14.1. *Possible techniques for the Yackel and Whealtey (1990) activity*

Based on this activity, we explored the didactic variables on which we could play in order to make the activity more complex or to generate different techniques for the pupils (Table 14.2). These didactic variables make it possible to vary the complexity of the activity and to make it a progression, in order to get pupils to use their SK, as well as to develop it.

The variables related to the images make it possible to identify, during the sharing phase, the elements that can make the reproduction of the figure more or less easy to do. In addition, this makes it possible to identify certain geometric properties that may be useful to observe in order to facilitate reproduction, such as symmetry, for example.

The variations in relation to the tasks make it possible to increase the complexity, as well as to vary the way of validating or "seeing" the figure. Indeed, the validation by description or the reproduction of the Tangram, only from a description of the figure without having seen it, allows us to use the IM of the figures in order to recreate the sought image. Moreover, the description of the figure to reproduce it without having seen it, requires the development of a precise vocabulary and a good articulation between the figures in order to ensure a faithful and most effective reproduction possible.

	Examples
Didactic variables for the images	– Color of the Tangram pieces used as pictures (identical or not). – Color of the pieces of the pupils' Tangram. – Relationship between the dimensions of the projected image and the pupils' Tangram. – Number of pieces. – 2D versus 3D.
Didactic variables related to the spatial and geometric properties of figures	– Line (present, partial or absent). – Symmetrical or not. – Isometric or not. – Compact, spread out or "perforated". – Orientation of the parts (similar or different). – Known images – unknown. – Alignment of the parts between them present or not.
Task variables	– Whether or not the number of pieces is mentioned. – Mention or not of the number of times it is presented. – Validation by a description. – Reproduction from a description only.

Table 14.2. *Didactic variables that can be used for this activity*

According to the guidelines set with the AGS, this activity addresses "organization" through the creation of MIs, since pupils cannot manipulate their Tangram pieces when they observe it and must, afterwards, reproduce it. The MIs are static and reproducible, given that they refer to the observed image. Articulation is also solicited since the pupils must take into account the position of each of the figures, their position in relation to the others and their alignment. It requires the pupils to have a mental representation of the observed image, entirely or partially, since the image to be constructed is no longer visible when reproduced. It takes place in micro space, since small objects representing geometric figures (triangle, square and parallelogram) are placed on the desk.

The tasks are essentially the observation of the image, the reproduction of this image using the Tangram and the verbalization of spatial reasoning. Even if the Tangram pieces are in 3D, they refer to geometric figures (2D). The questioning specifically concerns the creation of the MIs (see above). Three choices seem to be decisive, in this activity, for the development of SKs: that of forbidding manipulation during observation, which places the pupils in a position of anticipation (Braconne-Michoux and Marchand 2021); that of hiding the image to be reproduced during its construction, which provokes the need to resort to MIs; that of explicitly questioning the pupils on the creation of MIs ("What did you see in your head?") in order to make them aware of their own spatial reasoning and that of others.

14.4. Experiments with this activity in primary school and in teacher training in Quebec

This study, which falls mainly within the field of qualitative research, brings into play a broader didactic question: What theoretical foundations can allow the elaboration and analysis of teaching sequences that enhance the development of SK?

By analyzing some 20 very diverse teaching sequences, which were developed in the course of our work, it was possible to group them into three different profiles: *Quick Image*-type teaching sequences; teaching sequences dealing specifically with the development of SK and teaching sequences targeting related concepts or domains, but using SK. The sequences targeted by this chapter, developed from the work of Yackel and Wealthey (1990), are of the *Quick Image* type. In the following sections, we describe the methodology used for each of the two experiments, in primary school and in teacher training.

14.4.1. *Teaching sequence experimented in primary school*

The project in which the design and experimentation of this teaching sequence are part of has several characteristics of a collaborative research (Desgagné et al. 2001). It involved a researcher and nine primary school teachers on a question related to teaching practice (How can we improve the teaching and learning of SK at the primary school level?) and a didactic question (What theoretical foundations can enable the development and analysis of teaching sequences that enhance the development of SK?) It implies a co-construction approach in the elaboration of classroom activities that results from a mediation work between the two partners. This mediation is in line with the studies of the socioconstructivist perspective of knowledge and its development, where teachers are considered as reflective practitioners.

Of the teaching sequences developed in our work, three are detailed and analyzed in this chapter. We focus on five of the nine classes for which we also conducted data collection at different times. We targeted three lower primary school classes (ages 6–8) and two upper primary school classes, specifically sixth grade (ages 11–12), including a control group. For the three lower grades, semi-structured interviews were conducted with pupils at the beginning and at the end of the year. For the two sixth-grade classes, we employed the assessment tool developed by Ramful et al. (2017) at the end of the year. In addition, we felt it was relevant to include some empirical data that could illustrate the potential that these appear to have on the development of SK. Finally, comments made by the teachers of these primary school classes during this project, using our logbook, will enrich or nuance the results obtained. These three sequences are composed of a single activity (Yackel and Wheatley 1990), repeated several times throughout the school year, for which a learning progression was envisaged.

The sequence for the 1st cycle of primary school (6–7 years old) begins with Tangram composed of two figures. Each figure is a different color and the figures needed to reproduce the Tangram are mentioned to the pupils beforehand. The position of the figures is first prototypical and then non-prototypical; similar from one figure to another and then different from one figure to another. The relationship between figures is at first simple and then complex (unaligned, compact or "spread out"). And finally, Tangram and the figures that make them are firstly symmetrical and then asymmetrical (especially with the use of the parallelogram). Figure 14.3 shows two examples of Tangram from this sequence: the first is at the beginning of the sequence; the second at the end.

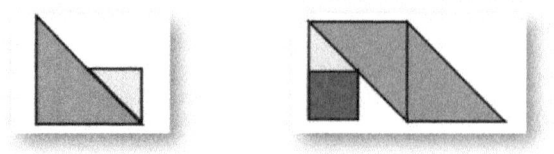

Figure 14.3. *Tangram from the 1st cycle teaching sequence (6–7 years). For a color version of this figure, see www.iste.co.uk/guille/tangible.zip*

The sequence for the 2nd cycle of primary school (8–9 years old) begins with a Tangram composed of four figures. All the figures are the same color, and the figures needed to reproduce the Tangram are mentioned to the pupils beforehand. The progression follows the same logic as the previous one. In addition, some of the lines between the figures in the Tangram can be hidden. Figure 14.4 shows two examples of Tangram from this sequence. The first is done at the beginning of the sequence, while the second is done at the end of the sequence.

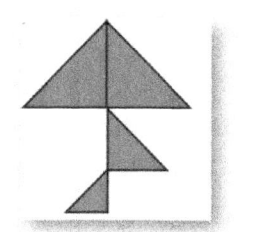

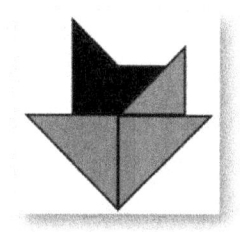

Figure 14.4. *Tangram from the cycle 2 teaching sequence (ages 8–9). For a color version of this figure, see www.iste.co.uk/guille/tangible.zip*

The sequence for the 3rd cycle of primary school (10–11 years old) no longer involves figures but objects like those found in the game *Architecto*[4]. It begins with compositions of four objects. All the objects are of the same color, and the objects needed for reproduction are mentioned to the pupils beforehand. As in the previous sequences, the position of the objects is first typical and then atypical, uniform and then variable from one object to another. The relationship between the objects is at first simple and then complex (e.g. not aligned, compact or spread out) and the objects are "simple" (like the prism) and then less familiar (like the bridge).

Moreover, the position of the objects can hide the determining faces that compose it or not. The determining faces which constitute it can be visible or hidden (e.g. the triangle of the prism with a triangular base). Finally, parts of objects may or may not be hidden by another object. Figure 14.5 shows two examples of a composition taken from this sequence, one done at the beginning, the other at the end of the sequence. These methodological elements express the main variables taken into account for the experimentation of this teaching sequence in primary school. Let us now look at its transposition to teacher training.

Figure 14.5. *Compositions of objects from the 3rd cycle teaching sequence (10–11 years)*

---

4 Lyons, M. and Lyons, R. (2005). *Architecto*. Brain Builder Series. FoxMind Games (ed.), Montréal.

14.4.2. *Teaching sequence tested in teacher training*

The methodology for this part of our work remains exploratory at this time. We have been conducting this teaching sequence for over five years, with approximately 90 pupils annually; however, we are only just beginning to study it. Our intention as trainers is twofold: to make students aware of the importance and contribution of this type of activity for primary and secondary pupils, and at the same time, to generate an opportunity to develop their own SK as future teachers. In order to inform our reflections on the contribution of this teaching sequence for teacher education, we employed the evaluation tool developed by Ramful et al. (2017), at the beginning and end of the session.

Therefore, the teaching sequence experimented with in teacher training at the Université de Sherbrooke was developed by the authors of this chapter, who offer a course on the didactics of geometry and measurement. The integration of the Tangram activity in almost all of the course sessions, a dozen in total, was carried out with the objective of having future teachers experience an activity that develops SK, as well as to explore the didactic potential of this activity and to observe the possible improvement of SK in adults.

The proposed sequence starts with a seven-piece Tangram in order to establish the students' SK. Thereafter, 2 four-piece Tangram identified beforehand, with the possibility of four observations, are performed. In the next session, a progression is established using different Tangram in four pieces. Over the course of the sessions, the number of pieces increases until it reaches seven, the number of observations is reduced to three, the pieces are no longer mentioned beforehand, and some traces are hidden. Validation is done with a description of the Tangram and no longer by its observation. In the penultimate session, the Tangram are described instead of being seen, in order to be reproduced. Figure 14.6 shows two examples of the sequence. The first is done at the beginning of the sequence, and the second at the end of the sequence, according to the methods just described.

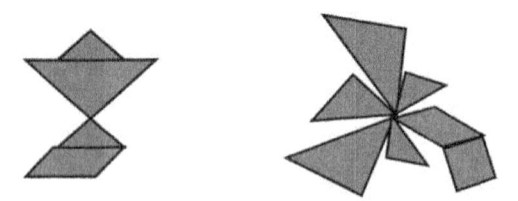

Figure 14.6. *Tangram from the teacher training sequence*

In parallel to this teaching sequence, we administered the Ramful et al. (2017) test, before and after the experiment. In addition, this test was also administered to 14 students enrolled in the "Sciences et technologies" degree in France. They therefore have a different profile from those enrolled in teacher training at our institution. The latter represent a group of French students with a scientific background who want to become school teachers. We wanted to study whether their profile in terms of SK was similar or divergent, given that their academic background was not the same. Indeed, the students of the bachelor's degree in social and educational adaptation do not have a scientific background, the majority of them have a background in the humanities, and many of them finished their training in Mathematics and Science at the end of their secondary education (16–17 years old).

14.5. Experiment results

The results related to the primary school experiment are presented first, after which we then present the results of the teacher training experiment in Quebec.

14.5.1. *Experiment results of the teaching sequence in primary school*

The teachers who participated in this project easily integrate this teaching sequence into their weekly routine, and it is a sequence that persists over time since the teachers are still using it today, despite the fact that the project has been finished for four years now. It is stimulating for the pupils, and the teachers are seeing improvements in pupil performance. For example, they note an improvement in their pupils' spatial and geometric vocabulary, a greater appropriation of the geometric priorities of simple figures, and an improvement in their SK. Table 14.3 shows some of the results obtained during the interviews, both a priori and a posteriori.

At the beginning of the year, it is possible to observe that 34% of the pupils were able to reproduce a picture with two triangles, and 57% with two triangles and a square positioned in an atypical way.

At the end of the year, the success rate for these two types of images is 100%. Moreover, 89% of the pupils succeeded with complex images made up of four figures, all positioned differently, and corresponding to the most complex images of the teaching sequence. Finally, 17% of the pupils demonstrated that they could go beyond the images proposed to them in class with an image composed of all the pieces (seven) of the Tangram illustrating a certain symmetry and including triangles positioned differently within the space, various alignments between the figures and a parallelogram with no axis of symmetry. These results highlight the malleability of MI creation and the level of complexity that can be exploited with pupils aged 6–8 years.

	Tangram present	Success rate
INTERVIEW *A PRIORI*		24/70 – **34%**
		40/70 – **57%**
INTERVIEW *A POSTERIORI*		70/70 – **100%**
		70/70 – **100%**
		62/70 – **89%**
		12/70 – **17%**

Table 14.3. *Pupil success rate a priori and a posteriori for 1, 2, 3 … imagine!*

We followed these pupils who had participated in the previous interview and various activities related to our work during the course of their 1st and subsequent years of their school career, and then re-collected data when they were in 6th grade. We administered the Ramful et al. (2017) test to these pupils as well as to a control group. This assessment consists of 30 questions: 10 are related to spatial orientation tasks, such as those defined above, 10 propose a mental rotation of an object and 10 are related to the transformation of a more complex object or set of objects (spatial visualization). Table 14.4 shows the average success rate of 11–12-year-old pupils for each of the three categories of questions. This average is obtained by adding the success rate of these pupils for each of the 10 questions in that category.

Pupils in the experimental group performed better on questions related to spatial orientation than those from the control group, with a difference of 9.2%. The difference between pupils in the control group and the experimental group for questions related to mental rotation is almost zero.

	Control group	Experimental group	Difference between the two averages
Spatial orientation	76.5%	85.7%	+9.2%
Mental rotation	66.2%	65%	-1.2%
Spatial visualization	49.6%	65.7%	+16.1%

Table 14.4. *Sixth-grade average test pass rates for each category of the Ramful et al.*

However, the most striking result was for the spatial visualization questions, where there was a 16.1% difference between the two groups. This category of questions was the least successful of the three categories for both groups. According to the framework, questions in this category refer to the "scenographic" level of abstraction represented by the highest level of abstraction. It is not surprising, therefore, that they are among the most complex. This gap is an average, but the gap between these two groups was very high for two questions in particular. It was 52.2% (19.2% vs. 71.4%) for the question asking to associate the cube with an icon per side (e.g. star) shown in perspective with its development, and 32.9% (38.5% vs. 71.4%) for the question asking to anticipate the shape of the section of the cube that was shown in perspective with the beginning of the cut illustrated with a knife and dotted lines.

Overall, we can see from this analysis that the SK can be dealt with at the primary school level, but that these are still under construction when pupils enter secondary school. It highlights that it is the SK related to spatial visualization requiring a level of scenographic abstraction that remains challenging for 11–12-year-old pupils. It is for these SK of the "visualization" type that we observe a greater gap in the level of performance between the control group and the group that was able to benefit from a teaching practice that valued the development of SK, through various teaching sequences including *1, 2, 3... imagine!*

14.5.2. *Experiment results of this teaching sequence in teacher training*

The initial results presented are related to the level of success on the first three Tangram of the semester, based on the number of observations needed to succeed. Table 14.5 presents the results for the 49 future teachers in the target cohort. No students were successful with the first Tangram on the first try, while 16/49 succeeded it after four observations. The second Tangram, consisting of four pieces, was achieved by three students on the first observation and by 48/49 after four observations. The third Tangram, composed of four figures in different orientations and more asymmetrically, was achieved by five students after the first observation but by 44/49 after four observations. We do not have the downstream picture of the teaching sequence for this cohort, but from an empirical standpoint, it is possible to say that they develop SKs in these future teachers.

	Number of observations				
	1	2	3	4	Failure
(rabbit figure)	0	1	12	3	33
(bird figure)	3	21	24	4	1
(cat figure)	5	18	21	4	5

Table 14.5. *Number of observations needed for pupils to reproduce the Tangram during the first session of 1, 2, 3… imagine!*

	Cohort A Before (n = 66)	Cohort A After (n = 62)	Cohort B Without (n = 14)
Spatial orientation	92.9%	94.9% (+2%)	89.3%
Mental rotation	75%	78.7% (+3.7%)	81.4%
Spatial visualization	66.7%	72.7% (+6%)	70%

Table 14.6. *Success rates of teacher education students on the Ramful et al. (2017) test*

In fact, during the last session, the three Tangram, composed of seven pieces, were successfully completed by the vast majority of future teachers. Moreover, several students told us that they had noted a certain progression for these SK. These results remain exploratory, but they also seem promising for teacher education. Table 14.6 outlines the Ramful et al. (2017) test results with this cohort, as well as with a cohort with a science academic background and in teacher training.

Recall that Cohort A experienced the *1, 2, 3... imagine!* teaching sequence described above and that this test was conducted before and after this sequence, and that Cohort B did not experience this sequence. If we compare the results of Cohort A with those obtained with the primary school pupils, it is possible to note three constants: the three categories of questions are in the same order of difficulty (orientation, rotation and visualization), the questions relating to spatial visualization remain the most complex and the greatest difference between the test administered before and after the teaching sequence is also the one related to spatial visualization.

Comparing the results for Cohort A in teacher education and Cohort B for those with a science background prior to teacher education, we might have expected a larger gap in terms of success on this test. The results illustrate little distinction. For spatial orientation, Cohort B scored lower than Cohort A in the pre-training phase, and the gap widened in the post-training phase, from 2.6% to 5.6%. For the questions related to mental rotations, Cohort B had a higher success rate than Cohort A before the sequence (6.4%), but this gap, which favored Cohort B, diminished with the results of Cohort A after the sequence (1.7%). Finally, for questions involving spatial visualization, Cohort B is slightly stronger (3.3% gap) when compared to Cohort A before the sequence, but Cohort A outperforms Cohort B once they have completed the teaching sequence (2.7% gap).

These observations, which remain exploratory at this stage, nevertheless seem to converge towards our working hypothesis that this type of teaching sequence has the potential to develop SK in all learners.

14.6. Conclusion

This empirical study illustrates the potential and versatility of this activity that we have entitled *1, 2, 3... imagine!* for the development of SK.

The alternation between the design of the AGS and the experimentation of teaching sequences, such as the one presented in this chapter, brings to the forefront the contribution of these two phases of our work, since one feeds into the other in a reciprocal manner. The AGS allows researchers to deepen their foundations and its use with teachers before and after the experimentation, providing them with guidelines to enrich their teaching sessions.

Moreover, still from an empirical point of view, it was possible to observe that the AGS allowed teachers to imagine a progression of activities that they could propose, in the short or medium term, to their pupils, by playing on the different didactic variables, and even in the longer term by articulating from one school year to the next. By exploiting the AGS, the aim is not only to produce teaching sequences but also to identify their structure, not only to solicit the SK necessary for the learning of mathematics, and other school subjects such as science, but above all to provide guidelines making their development possible for all, by proposing a management of didactic variables valuing an articulation and an interaction between the different levels of abstraction. The interest in pursuing and deepening our research in this sense seems relevant according to these initial data sets.

This type of didactic questioning in relation to the development of SKs at school is currently gaining momentum in the educational research community. We are currently at the stage of empirical study and, consequently, other research will be necessary to advance our knowledge – those valuing this development in both primary and secondary classes – in Mathematics as well as more broadly in Science or Technology.

14.7. References

Battista, M.T. (2007). The development of geometric and spatial thinking. In *Second Handbook of Research on Mathematics Teaching and Learning*, Lester, Jr. F.K. (ed.). National Council of Teachers of Mathematics, Reston, 843–908.

Battista, M.T. (2008). Development of the shape makers geometry microworld. In *Research on Technology and the Teaching and Learning of Mathematics: Vol. 2, Cases and Perspectives*, Blume, G.W. and Heid, M.K. (eds). Information Age, Charlotte, 131–156.

Berthelot, R. and Salin, M.-H. (1992). L'enseignement de l'espace et de la géométrie dans la scolarité obligatoire. PhD Thesis, Université de Bordeaux I.

Berthelot, R. and Salin, M.-H. (2000). L'enseignement de l'espace à l'école primaire. *Grand N*, 65, 37–59.

Braconne-Michoux, A. (2013). Des tours de D. Valentin aux représentations planes des objets de l'espace. In *Acte du colloque de la Commission Permanente des IREM sur l'Enseignement Élémentaire (COPIRELEM)* [Cédérom]. IREM de Brest.

Braconne-Michoux, A. and Marchand, P. (2021). La géométrie dans l'espace : une piste d'interventions auprès des élèves en difficulté ? In *La recherche en didactique des mathématiques et les élèves en difficulté : quels enjeux et quelles perspectives ?* Marchand, P., Adihou, A., Koudogbo, J., Gauthier, D., Bisson, C. (eds). Les éditions JFD, Montreal.

Brousseau, G. (1983). Étude de questions d'enseignement. Un exemple : la géométrie. *Séminaire de didactique des mathématiques et de l'informatique.* LSD IMAG and Université Joseph Fourier, Grenoble, 45.

Brousseau, G. (2001). Les propriétés didactiques de la géométrie élémentaire. L'étude de l'espace et de la géométrie. *Actes du séminaire de didactique des mathématiques.* Université de Crète, Rethymnon [Online]. Available at: https://hal.archives-ouvertes.fr/hal-00515110 [Accessed 27 September 2021].

Chastenay, P. (2015). From Geocentrism to Allocentrism: Teaching the phases of the moon in a digital full-dome planetarium. *Research in Science Education*, 46, 43–77 [Online]. Available at: http://doi.org/10.1007/s11165-015-9460-3 [Accessed 27 September 2021].

Cheng, Y.-L. and Mix, K.S. (2012). Spatial training improves children's mathematics ability. *Journal of Cognition and Development*, 15(1), 2–11.

Clements, D.H. (1999). Geometric and spatial thinking in young children. In *Mathematics in the Early Years*, Copley, J.V. (ed.). National Association for the Education of Young Children, WA.

Clements, D.H. and Sarama, J. (2011). Early childhood teacher education: The case of geometry. *Journal of Mathematics Teacher Education*, 14(2), 133–148.

Davis, B. and The Spatial Reasoning Study Group (2015). *Spatial Reasoning in the Early Years: Principles, Assertions, and Speculations.* Routledge, New York.

Desgagné, S., Bednarz, N., Couture, C., Poirier, L., Lebuis, P. (2001). L'approche collaborative de recherche en éducation : un rapport nouveau à établir entre recherche et formation. *Revue des sciences de l'éducation*, 27(1), 33–64.

Fujita, T. and Jones, K. (2007). Learner's understanding of the definitions and hierachical classification of quadrilaterals: Towards a theoretical framing. *Research in Mathematics Education*, 9(1), 3–20.

van Garderen, D. (2006). Spatial visualization, visual imagery, and mathematical problem solving of pupils with varying abilities. *Journal of Learning Disabilities*, 39(6), 496–506.

Gaulin, C. (1985). The need for emphasizing various graphical representations of 3-dimensional shapes and relations. *Proceedings of the 9th International Conference of the Psychology of Mathematics Education (PME)*, Utrecht, 53–71.

Gouvernement du Québec (2006). Programme de formation de l'école québécoise [Online]. Available at: http://www.education.gouv.qc.ca/fileadmin/site_web/documents/dpse/formation_jeunes/prform2001.pdf [Accessed 27 September 2021].

Hawes, Z., Lefevre, J.A., Xu, C., Bruce, C. (2015). Mental rotation with tangible tree-dimensional objects: A new measure tangible to developmental differences in 4- to 8-year-old children. *Mind, Brain and Education*, 9(1), 10–18.

Hegarty, M. and Kozhevnikov, M. (1999). Types of visual-spatial representations and mathematical problem solving. *Journal of Educational Psychology*, 91(4), 684–689.

Kahane, J.-P. (2002). Rapports et documents de synthèse de la Commission de réflexion sur l'enseignement des mathématiques. Rapport d'étape sur la géométrie et son enseignement [Online]. Available at: https://irem.u-paris.fr/documents-rapports-et-annexes-de-la-commission-kahane [Accessed 29 September 2021].

Kalogirou, P. and Gagatsis, A. (2012) Visualization as a factor of spatial ability and the relationship between pupils' spatial ability and geometrical figure apprehension. *Actes de colloque de l'espace de travail mathématique (ETM) au troisième symposium*. Montreal, 421–440.

Kyttälä, M. and Lehto, J. (2008). Some factors underlying mathematical performance: The role of visuospatial working memory and non-verbal intelligence. *European Journal of Psychology of Education*, 22(1), 77–94.

Marchand, P. (2006a). *La géométrie, tout un sport !* Éditions Bande Didactique, Collection Mathèse, Montreal.

Marchand, P. (2006b). Comment développer les IM reliées à l'apprentissage de l'espace en trois dimensions ? *Annales de didactique et des sciences cognitives*, 11, 103–121.

Marchand, P. (2009). L'enseignement du sens spatial au secondaire : analyse de deux leçons de troisième secondaire. *Canadian Journal of Science, Mathematics and Technology Education*, 9(1), 30–49.

Marchand, P. (2020). Quelques assises pour valoriser le développement des connaissances spatiales à l'école primaire. *Recherches en didactique des mathématiques*, 40(2), 1–44.

Marchand, P. and Braconne-Michoux, A. (2014). Quels types d'activités permettent de développer les connaissances spatiales chez les élèves du primaire : le cas de la boîte à image. *Acte du colloque COPIRELEM*. IREM de Nantes, Nantes.

Marchand, P. and Munier, V. (2021). Un levier pour une meilleure appréhension de l'espace en classe de sciences à l'école. *Canadian Journal of Science, Mathematics and Technology Education*, 21, 321–338.

Marchand, P. and Sinclair, N. (2017). Role of spatial reasoning in mathematics. In *Proceedings of the 2016 Annual Meeting of the Canadian Mathematics Education Study Group*, Oesterle, S. and Allen, D. (eds). Canadian Mathematics Education Study Group, 177–192.

MENJS (2020). Programmes d'enseignement – Cycle des apprentissages fondamentaux (cycle 2), cycle de consolidation (cycle 3) et cycle des approfondissements (cycle 4) Arrêté du 17-7-2020 and J.O. du 28-7-2020.

Mithalal, J. (2014). Voir dans l'espace : est-ce si simple ? *Petit x*, 96, 51–73.

Moss, J., Bruce, C.D., Caswell, B., Flynn, T., Hawes, Z. (2016). *Taking Shape. Activities to Develop Geometric and Spatial Thinking*. Grades K-2. Pearson, Toronto.

Mulligan, J., Woolcott, G., Mitchelmore, M., Busatto, S., Lai, J., Davis, B. (2020). Evaluating the impact of a spatial reasoning mathematics program (SRMP) intervention in the primary school. *Mathematics Education Research Journal*, 32, 285–305 [Online]. Available at: https://doi.org/10.1007/s13394-020-00324-z [Accessed 29 September 2021].

Munier, V., Marchand, P., Merle, H. (2011). L'enseignement de l'espace dans deux contextes différents d'apprentissage: En classes de sciences et de mathématiques. *Actes des 3e Rencontres Scientifiques Montpellier – Sherbrooke*, 159–173.

National Council of Teachers of Mathematics (NCTM) (2005). *Principles and Standards for School Mathematics*. National Academies Press, WA.

National Council of Teachers of Mathematics (NCTM) (2006). *Learning to Think Spatially: GIS as a Support System in the K-12 Curriculum*. National Academies Press, WA.

Ontario Ministry of Education (2014). Paying attention to spatial reasoning support document for paying attention to mathematics education [Online]. Available at: http://www.edu.gov.on.ca/eng/literacynumeracy/lnspayingattention.pdf [Accessed 27 September 2021].

Panaoura, G., Gagatsis, A., Lemonides, C. (2007). Spatial abilities in relation to performance in geometry tasks. In *Proceedings of the Fifth Congress of the European Society for Research in Mathematics Education*, Department of Education University of Cyprus (ed.). University of Cyprus, Larnaca.

Parzysz, B. (1988) "Knowing" vs "seeing". Problems of the plane representation of space geometry. *Educational Studies in Mathematics*, 19(1), 79–92.

Perrin-Glorian, M.-J. and Godin, M. (2018). Géométrie plane : pour une approche cohérente du début de l'école à la fin du collège. *Actes du colloque CORFEM* [Online]. Available at: https://hal.archives-ouvertes.fr/hal-01660837v2.

Piaget, J. and Inhelder, B. (1977). *La représentation de l'espace chez l'enfant*, 3rd edition. Presses universitaires de France, Paris.

Ramful, A., Lowrie, T., Logan, T. (2017). Measurement of spatial ability: Construction and validation of the spatial reasoning instrument for middle school pupils. *Journal of Psychoeducational Assessment*, 35(7), 1–19.

Salin, M.-H. (2008). Enseignement et apprentissage de la géométrie à l'école primaire et au début du collège : le facteur temps. *Bulletin de l'APMEP*, 478, 647–670.

Salin, M.-H. (2013). Quelques remarques autour des finalités de l'enseignement de la géométrie à l'école primaire. *Actes du XLème colloque COPIRELEM, Enseignement de la géométrie à l'école*. IREM et IUFM des Pays de la Loire.

Shumway, J.F. (2013). Building bridges to spatial reasoning. *Teaching Children Mathematics*, 20(1), 44–51.

Sorby, S.A. and Panther, G.C. (2020). Is the key to better PISA math scores improving spatial skills? *Mathematics Education Research Journal*, 32, 213–233 [Online]. Available at: https://doi.org/10.1007/s13394-020-00328-9.

Tolar, T.D., Lederberg, A.R., Fletcher, J.M. (2009). A structural model of algebra achievement: Computational fluency and spatial visualization as mediators of the effect of working memory on algebra achievement. *Educational Psychology*, 29(2), 239–266.

Uttal, D.H., Meadow, N.G., Tipton, E., Hand, L.L., Alden, A.R., Warren, C., Newcombe, N.S. (2013). The malleability of spatial skills: A meta-analysis of training studies. *Psychological Bulletin*, 139(2), 352–402.

Verdine, B.N., Golinkoff, R.M., Hirsh-Pasek, K., Newcombe, N. (2017). Links between spatial and mathematical skills across the preschool years. *Monographs of the Society for Research in Child Development*, 82(1), 77–85.

Wai, J., Lubinski, D., Benbow, C.P. (2009). Spatial ability for STEM domains: Aligning over 50 years of cumulative psychological knowledge solidifies its importance. *Journal of Educational Psychology*, 101(4), 817–835.

Yackel, E. and Wheatley, G.H. (1990). Promoting visual imagery in young pupils. *Arithmetic Teacher*, 37(6), 52–58.

15

What Use of Analysis a priori by Pre-Service Teachers in Space Structuring Activities?

15.1. Introduction – an institutional challenge of transposing didactic knowledge

Among the professional gestures expected of teachers is the ability to anticipate the different procedures that can be used by pupils and to identify the knowledge contained in the tasks proposed to them. It is undoubtedly with regard to these aims that a priori analysis figures in the content of most Francophone teacher training courses. However, beyond mere planning and anticipation, the analysis of actions such as intervention (scaffolding) with pupils is also expected. This requires, through training, the construction of systems that suggest an implicit recourse to a priori analysis can be mobilized for this. It is the strength of this implication and the impact of the underlying disciplinary knowledge that we propose to study here. Therefore, after conducting a study on the transposition of the a priori analysis as a professional gesture, we will question the mobilization of the latter by young teachers in French-speaking Switzerland, in training within the framework of an activity aimed at structuring space for young pupils (3–4 years old).

15.1.1. *Choice of external transposition: institutional constraints*

In Switzerland, education and the school system as a whole are a cantonal responsibility. The same is true for teacher training. Each (cantonal) training institute is free to choose its own training content, provided that it respects very

---

Chapter written by Ismaïl MILI.

broad constraints resulting from inter-cantonal agreements. Therefore, it is the trainers themselves, with regard to institutional constraints and their conceptions of the teaching profession, who take charge of the external transposition of the didactic knowledge that will be worked upon in training.

15.1.1.1. Constraints inherent to the reference works

Although these inter-cantonal agreements do not propose standardized study or training plans, they nevertheless lead to the common development of unique pedagogical resources (*Moyens d'Enseignement Romands*, MER), the use of which becomes de facto mandatory for all teachers and schooling. With regard to mathematics resources in elementary school, the editors of the latest edition of the MER, introduced in 2018, advocate in the general comments accompanying the activities an explicit use of a priori analysis:

> It will therefore be necessary to determine the conditions for this transition [in this case, an informational leap] to be made indispensable. To do this, it will be necessary to carry out an *a priori* analysis of the activities proposed to the pupils (see text on *a priori* analysis) and identify the choices to be made (didactic variables) to force this transition (MER, n.d. extracted from Le Nombre – Premier Apprentissage – Cycle 1, *translated by author*)

Of course, the compulsory nature of the use of these resources will have an impact on the choices of external transposition for the content of different training institutes: these will have, from then on, an almost explicit mandate to teach a priori analysis, questioning the nature of this *knowledge to be taught* and of the possible distinctions to be made with a *scholarly knowledge*, which would remain to be determined[1].

15.1.1.2. Constraints inherent to the training institute

Like several Swiss institutes, the Haute Ecole Pédagogique du Valais (HEP VS) chose to train (and evaluate) teachers by professional competencies in 2013, which necessitated a certain "linking" of course and practicum content in the pupil's training path.

Among the five Francophone training institutes, our interest in the HEP VS is motivated by the explicit appearance of a priori analysis in the practicum criteria (the Francophone version of which is reproduced in Table 15.1): it is therefore

---

[1] The bibliography presented by the MERs does not provide references or an understanding of the term "analysis a priori".

possible to infer that it must be included in the content of the various courses offered[2].

Specific criteria	Evaluation
SC 1	Clarify the object of study through the analyses (conceptual and a priori) and delimit it in order to promote its accessibility (identification of difficulties in making choices).
SC 2	Design and plan in writing a teaching/learning device taking into account the points raised in the analyses (of conditions, conceptual, a priori).
SC 3	Accompany pupils in their learning. Justify his/her interventions in light of teaching/learning theories.
SC 4	On the basis of the planned objective, anticipate and set up a formative evaluation system: analyze the pupils' oral or written traces in order to propose regulations to support learning.

Table 15.1. *Evaluation criteria for the 3rd-semester internship*

The titles of the different criteria tell us that, in addition to planning (SC 1 and SC 2), future teachers will be evaluated – during their third-semester internship of a training program that includes six semesters – on the relevance of their interventions and on their ability to justify them (SC 3). The same applies to the creation of a formative evaluation system (SC 4). Therefore, beyond a single planning technique, this a priori analysis, seen as knowledge to be taught in training, should also, despite its implicit nature in the formulation of SC 3 and SC 4, be able to be mobilized in order to justify the accompaniment of pupils and the setting up of a formative evaluation system. It is therefore up to the trainers to ensure the transposition (Chevallard 1985) of this didactic knowledge from Semester 3 onwards, both in terms of techniques and with the aim of bringing out its different applications, or to use the terms developed later, the *work of organization*, in the sense of Chevallard (2002). It is on the basis of this field of applications that we will circumscribe this a priori analysis constituting a *knowledge to be taught* – characterized as a professional gesture and which will be distinguished from that thematized by Artigue (1990a).

15.2. Theoretical framework

This first distinction, inherent to external transposition (Chevallard 1985), leads us to describe, in this section, the internal transposition of this knowledge. As

---

[2] At this institution, didactic courses in mathematics are located in Semesters 3 and 4.

indicated above, the didactic modules of mathematics are located in Semesters 3 and 4 of a six-semester program. With regard to institutional requirements (evaluation during Semester 3), the a priori analysis is addressed from the first sessions of the first module. The team of trainers in charge of these modules has chosen to deepen numerical notions in Semester 3 and geometrical/spatial notions in Semester 4.

15.2.1. *Choice of internal transposition: the moments of the study of the analysis a priori*

In the desire to inscribe within its training system a paradigm of questioning the world (Bosch et al. 2018, cited by Candy 2020), the team of trainers chose to simulate professional practice in training. It thus based itself on both a role-playing device (Lajoie and Pallascio 2001) and on "the moments of the study" (Chevallard 2002), considering the a priori analysis as knowledge to be taught generating a punctual Mathematical Organization[3]. This overlap generates some adaptations to the training device initially set up by Lajoie and Pallascio (2001). Therefore, for a task T, consisting of "carrying out an a priori analysis", we find in the planning of Semester 3 the following progression, foreseen to take place over approximately six 2-hour long sessions:

– *Moment of the (first) encounter with task T:* once the whole "role-playing" device has been presented (everyone knows that an intervention and a debate will follow), the pupils are led, in groups, to solve a problem in several different ways, without revealing their processes to the other groups.

– *Moment of the exploration of T and the emergence of technique t*: a pupil who will play the role of the teacher is randomly drawn from one of the groups, which have also been randomly selected. The other groups each send a representative to play a pupil. The "teacher" has to react to the productions presented to them (which they had not necessarily anticipated). A collective debate follows the intervention. During the debate, the trainer opts for a listening posture (without intervening) and leads the pupils to comment and justify the interventions they made[4].

– *Moment of the construction of the technological-theoretical block*: extraction from the trainer of the elements of debate that can constitute technological elements.

---

3 These elements are presented here in a very descriptive way and are not intended to justify the theoretical choices and analogies made by the trainers, which will undoubtedly become a topic for future publications.

4 During this stage, the students activate spontaneously, with regard to the position of the trainer, elements such as the importance of epistemological analysis in the context of numerical activities. This position is justified by the goal for trainers to fit into a paradigm of "questioning the world".

– *Moment of the institutionalization*: carried out by the trainer who leans on the elements of debate.

– *Moment of the work of the mathematical organization (and in particular of the technique)*: application on other types of situations, in particular for purposes of support and the creation of an evaluation device.

15.3. Research questions

Considering the succession of disciplinary contents dealt with over the course of the semesters, we will use these institutional configurations to verbalize the following research questions. Is a training device that recreates the terrain (role-playing) and the didactic indications proposed to the teachers in training by the various institutions sufficient for the appropriation and mobilization of a priori analysis (as defined at the end of the previous section) within the framework of space structuring activities? Still in such a training device (role-playing game), what are the characteristic elements of a priori analysis reinvested during the preparation and piloting of space structuring activities, by pupils who have previously worked during their training on the different applications of this professional gesture in a numerical domain? Is an epistemological analysis of the objects manipulated at the level of knowledge carried out by the students in training within a framework of space structuring? Can a notable difference be noted with activities of a numerical nature?

15.4. Methodology

In order to allow for the comparison aimed at through our second research question, this training device is based, at the beginning of Semester 3, on a numerical activity which will be described in the following section.

At the beginning of Semester 4, a similar device is reproduced with the same pupils (admittedly, without being able to speak then stricto sensu of a *first encounter*), this time based on an activity of structuring space.

Therefore, our body of data focuses each time on recorded discourse (debate described at the time of *T's exploration* – hereafter Phase 2) and written traces (produced at the time of the *first encounter* – hereafter Phase 1) gathered during two successive academic years (2018–2020), each cohort of future teachers comprising approximately 30 pupils.

As far as our first research question is concerned, we refer to Artigue (1990a) in order to characterize the possible mobilization of the a priori analysis (as defined in section 15.1.1), which can be observed with regard to the modulation, during the piloting of the activity (Phase 2), of the identified didactic variables (Phase 1) or of various mentions during the debate (Phase 2). The same applies to the second research question: it is also in these traces (Phases 1 and 2) that we extract possible questionings or epistemological studies relating to the hosted knowledge.

15.4.1. *Selection of activities and brief analysis*

The numerical activity is chosen (Semester 3), in view of the plurality of available resolution techniques – themselves resulting from various technologies – and the different ostensives that can be mobilized (correspondence table, linear equation, etc.). The trainers chose a situation (Figure 15.1) that stated the mobilization of the notion of proportionality[5].

> 5. The Politics of Small Steps
>
> To cover the 36 meters of the perimeter of her rectangular pool, whose length is twice the width, Paulette counted 54 steps.
>
> How many steps did it take her to cover its width alone?

Figure 15.1. *Excerpt from MER 8H (pupils aged 12)*

In order to test the possible (re)mobilization of the a priori analysis (see the first research question), it was then proposed to the pupils, during one of the first sessions of the following semester, the situation *Une Chaise pour deux* (A Chair for Two), extracted from the *Axe Thématique Espace des Moyens d'Enseignement* (1-2H, pupils of 4–5 years old):

– Time 1 (real-time imitation): when entering the room, the pupils, in pairs, orient their chairs in the same direction so that they can turn around without touching the other chairs; "You take turns making up a statue using the chair. The other imitates the statue and the first validates or corrects it".

– Time 2 (sculpture by instructions): three roles are proposed: the model, the sculptor and the sculpture (a screen separates the model from the sculpture). The model: "Invent a statue using the chair." Sculptor: "Tell the sculpture what to do to reproduce the model." To the sculpture: "Do what the sculptor tells you."

---

5 This is at least the objective attributed to it in the comments accompanying the statement. An a priori analysis, difficult to develop here, would however show that this can be discussed.

Located at the beginning of a teaching sequence, this activity, whose stated objective is to lead the pupil to "Determine his/her position or that of an object according to different reference points", is presented as an "introductory activity" with the intention of "helping the pupil to discover (construct) the knowledge taught". Its place upstream of a sequence and the intention announced by the authors of the MER, would justify that the teacher carries out, during the piloting, some modulations of the numerous didactic variables, all the more so given that the didactic environment thus constructed proves to be sufficiently antagonistic so as to issue to the "sculptor pupil" with strong feedback on his/her production.

One of the interests of this activity lies in the fact that, in addition to its statement, the didactic variables are not listed exhaustively by the MERs. Only those related to the types of indications provided or the number of consultations allowed are implicitly mentioned in the comments:

> The activity can be conducted without verbalization but in the form of modeling or sculpture. The sculptor must reproduce the proposed model by manipulating the sculpture. It is possible to limit the number of times the model is consulted. The validation is done by putting the model and the sculpture together. In this case, it is only at this point that verbalization is required (MER, n.d., comments on the "A Chair for Two" activity, *translated by author*).

Figure 15.2. *Training situation where the pupil playing the role of the teacher (scarf) intervenes with the pupil-sculptor (red sweater), in charge of giving instructions to the statue (striped sweater). For a color version of this figure, see www.iste.co.uk/guille/tangible.zip*

The spatial knowledge mobilized differs according to the respective orientation of the chairs – while satisfying the instruction, they can be arranged in line with each other (respecting the "same orientation" constraint) and thus promote elements inherent in translation and not in symmetry, as suggested by the arrangement shown in Figure 15.2 – or of the different protagonists.

Therefore, following the example of the numerical activity presented in the previous semester, several resolution procedures that call upon different types of identification can be mobilized (recourse to bodily laterality, internal and/or external identification of the situation, cardinal points, etc.), which can then generate questions of an epistemological nature as to the hosted knowledge.

15.5. Results

First of all, we note that, in the case of the numerical activity, even if the importance given to each of the parts varies, all the debates (Phase 2) present a relatively similar structure: they are initiated by questioning the validity and the status of the techniques mobilized ("But are we allowed to do that?", "Is it on the syllabus?", etc.) before questioning different epistemological components ("But what is proportionality apart from a [cross-reference] table?") and ending on the interest of considering the elements developed during the a priori analysis in order to build an evaluation device ("Ah, but that's how you would make your evaluation criteria?!?"), etc. In general, the conclusion of the three debates points to the need to carry out an epistemological analysis of the objects handled, in order to be able to pilot the activity a little more easily.

Finally, a significant proportion of the traces produced during the preparation (Phase 1) testifies to a search for elements of the statement which are, moreover, modified by the teachers during the various interventions (Phase 2). As an example, we point out a large panel of ostensives and numerical values.

On the other hand, in the case of the spatial activity, none of the traces produced during the preparation (Phase 1) seem to point to elements that can then be modulated by the teacher. All the pupils who took on the role of the teacher opted for an intervention centered on the validity of the production ("Do you have the impression that you have done it right?", "So, it's not right") and on the reproduction of the model's position ("If I put myself like this and if I turn around, do I have the same thing?"). In the vast majority of cases, the spatial vocabulary was not mobilized during the intervention (Phase 2), neither by the pupil nor by the teacher ("You put the other hand on the chair and you raise the other arm"). The various discussions (from which no common structure seems to emerge) do not propose any links between the preparation and the piloting; none of them raises the importance of an epistemological analysis.

15.6. Conclusion

We thus note several differences between numerical and spatial activities.

In the first case, while the a priori analysis is still unknown to them, the pupils tend to look for several procedures during the preparation, and leave themselves the latitude to modify various elements of the statement during their interventions. In a certain sense, an a priori analysis is carried out en acte by the group during the exchanges (Phase 2) at the end of which different applications are spontaneously pointed out (then taken up again later, see Phase 5). At a more advanced stage of their training, however, the trainee teachers have difficulty anticipating various procedures, identifying didactic variables and mobilizing them during the piloting of spatial activities. Moreover, no recourse is made to the various didactic concepts stabilized beforehand in the training, during the discussions on the spatial activity. It is the same for the epistemological analysis, however, carried out beforehand in a numerical framework and then considered necessary.

We hypothesize strongly that, unlike notions of proportionality, the self-evident nature of spatial notions prevents trainee teachers from "detaching themselves from the illusion of transparency" (as noted by Artigue 1990b, p. 245, *translated by author*).

Therefore, we conclude that, contrary to what seems to prevail in the context of numerical activities, the training device set up in the training does not allow trainee teachers to appropriate a priori analysis as a professional tool for space structuring activities, at least in meso-space: despite the numerous didactic variables that can be mobilized, it is not perceived as having an impact on scaffolding in the session or on professional interventions in general.

15.7. References

Artigue, M. (1990a). Ingénierie didactique. *Recherches en didactique des mathématiques*, 9(3), 281–308.

Artigue, M. (1990b). Epistémologie et Didactique. *Recherches en didactique des mathématiques*, 10(2.3), 241–286.

Candy, J. (2020). Etude de la transposition didactique du concept d'idéal : écologie des savoirs et problématique de l'entrée dans la pensée structuraliste, en France et en Suisse romande. Anneaux et algèbres. PhD Thesis, Université de Montpellier.

Chevallard, Y. (1985). *La transposition didactique : du savoir savant au savoir enseigné*. La Pensée Sauvage, Grenoble.

Chevallard, Y. (2002). Organiser l'étude 1. Structures & fonctions. *Actes de la XIe école d'été de didactique des mathématiques* (Corps, 21–30 August 2001). La Pensée Sauvage, Grenoble.

Lajoie, C. and Pallascio, R. (2001). Le jeu de rôles : une situation-problème en didactique des mathématiques pour le développement des compétences professionnelles. In *Actes du colloque GDM 2001*, Portugais, J. (ed.). Université de Montréal.

MER (s.d.). Moyens d'Enseignement Romands [Online]. Available at: http://www.ciip-esper.ch/#/discipline/5/1,2/ [Accessed 19 January 2020].

PART 4

Conclusion and Implications

16

Questions about the Graphic Space: What Objects? Which Operations?

In this chapter, we want to come back to some elements treated in the different chapters that deal with the semiotic dimension of geometric work. Then, we will ask some questions to open up some lines of thought about graphic space, the interface between tangible space and geometric space.

16.1. Semiotic tools of geometric work and graphic space

The semiotic dimension is intrinsically linked to mathematical activity and has been worked on by many researchers in mathematics didactics. Let us mention in particular Chevallard (1994) with the notions of ostensives and non-ostensives, Duval (1995) with the notion of register of semiotic representation, Radford (2014) with the theory of objectification, Arzarello (2006) with the notion of the semiotic bundle, and Bartolini-Bussi and Mariotti (2008) with the theory of semiotic mediation. Duval (1995) speaks of the cognitive paradox of mathematical thinking to indicate that "on the one hand, the apprehension of mathematical objects can only be a conceptual apprehension and on the other hand, it is only by means of semiotic representations that an activity on mathematical objects is possible" (op. cit., p. 38, *translated by author*).

Within the framework of the study of space and geometry, this semiotic dimension has also been worked on by several researchers, essentially in two interdependent ways: the first one around the language dimension in connection with situations of formulation or validation (among others, Bulf, Celi, Chesnais, Hache, Mathé, Mithalal, Perrin-Glorian), or with the geometric lexicon, description,

---

Chapter written by Teresa ASSUDE.

demonstration texts (Duval) or other kinds of texts like construction programs (Millon-Fauré et al. 2020); the second dimension is that of the figures, the relationship to the figures, their treatment and the conversion between different registers of representation. Duval's work has been interested in these questions for a very long time with the problem of visualization (iconic and non-iconic), of the different modes of apprehension of a figure (perceptive, operative, sequential, discursive), of the transformations of representations (processing, dimensional deconstruction, conversion). The question of the articulation of these dimensions has been raised and addressed by numerous works. For example, Mathé and Mithalal-Dozé (2019) indicate several stages of articulation between language and figure work:

> Moving from the instrumented analysis of drawings to the formulation of a geometric analysis of figures presupposes being able to: i. name the geometric objects mobilized, 0D, 1D, 2D figurative units of the complex drawing under study; ii. name the relations between these geometric objects independently of the instruments that carry them; iii. articulate mathematical discourse and the organization of the geometric analysis of the drawings; iv. move from a pragmatic description of a figure to an analysis that allows it to be characterized (defined) geometrically (op. cit., p. 69, *translated by author*).

The semiotic dimension is not restricted to the linguistic and graphic dimensions, but other dimensions can be taken into account, such as the gestural (body) or material register.

The semiotic question is addressed in this work. Céli's work (Chapter 3) is an example of the importance of the physical (body) and material dimension of using templates. Assude (Chapter 6) also addresses the importance of the perceptual-gestural and material level of the instrumental dimension, in the link to be made between the material gestures with the tool and perceiving the effects of these gestures. The question of the geometrical lexicon is important to define the geometrical objects. However, as Millon-Fauré et al. show (Chapter 13), teachers working with allophone pupils experiencing French language difficulties use vernacular language instead of the terms of the geometric lexicon to make certain terms understood in a sort of "didactic repression".

This step could be seen as an intermediate step, but it can also create obstacles. Would it not be better to go through stages such as schematization? Bloch and Pressiat (2009) insist on the role of schematization and semiotic tools as a means of modeling objects in the tangible space. For these authors, schematization plays a pivotal role in the passage between tangible space and geometric space and should be the object of particular attention in elementary school. These diagrams are indeed objects of graphic space even if they are not yet representations of geometric objects. This graphic space is the place of representations of material objects, of positions, of movements in and of the tangible space, as well as the place of representations of geometrical objects. On one hand, it is a question of object and action models, and on the other hand, "physical" representations of theoretical objects and relations.

In the continuity of the work by Duval and the Lille group (Duval et al. 2004) on figure restoration situations, Mathé and Perrin-Glorian (Chapter 1) approach what could be a "practical axiomatics" of a geometry of tracings, and show the evolution of these situations of action, into situations of formulation and validation by means of graphical and language work. The description of the figures is central because it "requires producing from this figure sufficient information in order to characterize it without ambiguity". In fact, these authors affirm that "the description of figures with a view to their construction and the validation processes that it generates seem to us to be a lever for the transition between material geometry and theoretical geometry". According to them, one of the difficulties facing pupils is how to move from the description of objects or actions to the description of relationships, in order to characterize geometric objects (naming and defining these objects). In addition, the construction programs are also presented as being able to ask for the information needed to construct the figure, as well as to ask the question: "How, from what we know (notably through the construction program), can we establish other properties using what we have?" (op. cit., *translated by author*). This type of questioning and figure analysis could get pupils into a proof process. The description of figures is also approached by Assude (Chapter 6), in the context of the reproduction of figures by articulating the work with a Dynamic Geometry software and with pencil and paper. In the situation presented, it is a question of describing objects in graphic space, of describing actions, as well as describing relations (of belonging or of relative positions between the figurative units). The author also indicates the importance of moving from the description of actions, or objects, to the description of relations. Finally, two other chapters likewise engage in this semiotic dimension: one by the coding of figures (Coutat, Chapter 4), the other by the freehand drawings (Vendeira-Maréchal, Chapter 5).

In this research, freehand drawing is a field for graphic experimentation in geometric shape recognition and classification activities. The succession of freehand drawings provides a means of controlling the environment and emphasizes the properties of the figures. Coding is also a means of condensing some of the information in the figures, including the properties of the figures. Coutat (Chapter 4) shows that the arrangement of several cards, where properties are coded, would allow for the description of (non-usual) figures without using speech. Coding is also a way of noting observations that can be derived from the tangible space or from the graphic space.

16.2. Graphic space: graphic expressions, denotation and meaning

The work of Perrin-Glorian and Godin (2018) shows the pivotal role of graphic space in the transition from tangible space to geometric space, with the "graphical space of representations" seen as an interface between these two spaces. In this graphic space, "one finds both 2D or 3D representations of objects or situations of the tangible world, and representations of geometric objects" (Mathé et al. 2020, *translated by author*). In the previous section, we highlighted the coexistence of different representations of tangible objects and/or geometric objects (figures as well as diagrams, freehand drawings, etc.) that can be assembled in what is called the graphic space. In this conclusion, we want to ask three questions that have implications for future research: How can we define this graphic space? What objects could belong to this graphic space? What do we do with these graphic objects?

16.2.1. *How can we define the graphic space?*

The graphic space (or graphic framework) is the set of objects that can be inscribed on a surface and that are visible (or palpable) traces on a surface. This definition makes it possible to put together very different objects. For example, writing is also part of the graphic objects, as shown by the work of the anthropologist Jack Goody (1979) on "graphic reason", but we are not going to be interested in what can be designated by written speech (texts, sentences, words). The objects of graphic space on which we focus here are those that can make links between the tangible space and the geometric space, other than speech, even if these are essential and necessary for working in geometric space: diagrams, drawings, figures, plans, graphs, sketches, images or any other expression that leaves a trace on a surface.

We will call them *graphical expressions* because they express something: displacements, positions, movements, geometrical objects, geometrical relations, actions, objects of the tangible space, etc. Using an analogy with the algebraic domain (even if the domains are different), just as algebraic expressions express something, in particular computational programs, graphical expressions express objects, actions of the real world and/or geometric objects, relations or properties of objects.

16.2.2. *Which objects in the graphic space?*

Mathé and Perrin-Glorian (Chapter 1) present graphic space as an "experimental ground for theoretical geometry" and speak of diverse interpretations or misunderstandings, stemming from the fact that graphic objects (in particular the "material figure") can be seen by themselves or represent geometric objects or objects of tangible space.

Indeed, a number of studies have shown the difficulty for pupils to identify the status of figures in either Geometry 1 or Geometry 2 (Houdement and Kuzniak 2006). Pupils' difficulties can be identified by the fact that they do not associate graphic expressions with what they express. Frege's distinction between meaning and denotation (used in the work on pupils' difficulties with algebra by Drouhard (1995) among others) could perhaps help us understand some of these difficulties. According to Frege (1892–1971), the denotation of an expression corresponds to the object that is designated, while the meaning is the "mode of donation" of the object. Taking an example in the numerical field, the denotation of the expression "7 + 1" is the number 8, whereas the meaning of this expression ("mode of donation") is that the number 8 is the successor of 7. Taking the expression "2 × 4" which denotes the same number, the meaning is not the same because the expression gives the idea that 8 is the double of 4. The two expressions are equivalent because they denote the same object (the number 8), but they are different (different scripts).

Could this distinction be applied to objects in graphic space, in particular to graphic expressions? What would be the relevance of doing so in this domain? These questions open up different avenues of work. The first would be to consider that such and such a graphic expression would denote such and such an object, and would show such and such a property of the object, which would not necessarily be the same thing for a graphic expression that would be equivalent to it, because it would denote the same object. Let us give an example.

When using geometry software, some pupils see a succession of drawings when they move one of the points of the initial figure: for them these are different drawings that refer to different objects. The interest in moving the figure consists of seeing what remains invariant in this entire class of drawings (the figure as a geometric object). If we construct a rectangle based on some of its properties, when we move the initial rectangle, we find different graphical expressions of the same denoted object, but these equivalent graphical expressions may show different properties: for example, we may see some of these graphical expressions as squares, leading to the identification of the square as a particular rectangle. Conversely, pupils might see that the denoted object is not what they originally thought, for example: thinking that the denoted object was a square because one of the graphical expressions was a "square" would be invalidated by dragging.

Would it not be necessary for pupils to distinguish the denoted object from the different graphical expressions that can show different properties of the denoted object? Again, in analogy with the results of research in the algebraic domain, we can put forward a working hypothesis: knowing what the graphic expressions denote would be an asset for geometric learning. This hypothesis should be verified, not only in relation to figures but also in relation to other graphic expressions, such as diagrams or freehand drawings. In relation to freehand drawings, some pupils have difficulty identifying the denoted object, because they focus on what the graphic expression shows, that is, such an angle is not straight even if the coding indicates it as straight.

16.2.3. *Graphic expressions: which operations?*

The interest of examining graphical expressions lies in the possibility of operating on these expressions, but what would be the possible operations with these expressions? This question, like the previous ones, is not new. Duval (2006) talks about it as follows:

> In contrast to this reduction of semiotic representations to the sole role of substitutes for mathematical objects, or to that of expressions of mental representations, we will focus on what constitutes the fundamental characteristic of any mathematical approach: the transformation of semiotic representations. For in mathematics, a representation is only interesting insofar as it can be transformed into another representation (op. cit., p. 74, *translated by author*).

Duval speaks here, in fact, of an operation that seems essential to mathematical work: the transformation of semiotic representations. For a very long time, Duval (1994, 1995, 2005) has been working on this question and has identified three cognitive activities for a semiotic system to become a register of representation: the formation of a representation, the processing of a representation in the same register and the conversion of a representation into a register other than the initial one. We will start from these three types of cognitive activities in order to highlight what corresponds to these cognitive operations in the graphic space, what we could call *graphic operations*. We are opening up potential avenues of research in continuation of Duval's work.

A first type of graphic operation could be the *composition of graphic expressions*. It consists of composing two or more graphic expressions to obtain a graphic expression. The starting graphical expressions can denote objects of the tangible space or geometrical objects. The composition of graphical expressions is a graphical expression which can have, or not, the same denotation. An example of a composition of graphical expressions can be taken from Coutat (Chapter 4). Code-maps are graphical expressions of properties. When we compose these graphic expressions, we obtain another graphic expression that denotes another object.

Another example can be taken from Assude (Chapter 6) with the situation "compositions" in which the pupils have to compose three basic figures to reproduce a given figure using pencil and paper. The composition leading to the complex figure implies, in fact, operations of identification of these figurative units, then of identification of the relative positions between these figurative units. The complex figure thus appears as the composition of graphic expressions (i.e. simple figurative units having given relations). This operation is determined by given rules, depending on the tools, the constraints and the type of problem to be solved.

The opposite operation of composition is the *decomposition of a graphic expression*. Decomposition consists of finding the elements and relationships that constitute it. In the case of a figure, it is a matter of analyzing the figure and finding the essential elements that constitute it. Many works on the analysis of figures to reproduce or restore a figure are placed within this framework. Let us look at another example of graphic expression: the plan of a classroom. The composition of a plan supposes that we place ourselves in a modeling problematic (Berthelot and Salin 1992). It is a matter of identifying variables: the shape of the room; fixed elements (doors, windows, blackboard, etc.); variable elements (chairs, desks, cupboards, etc.).

It is then necessary to establish relationships between these variables (the relative position of the fixed elements between them, the relative position of the variable elements with respect to the fixed elements, etc.). The composition of the plan consists of coding these elements and the relative positions of these elements. Coding is an essential element for composing the plan. This plan is a graphic expression of a tangible space and it shows certain relationships (arrangement of the desks between them, the teacher's desk and the blackboard, etc.). By placing ourselves at the level of cognitive operations, we are here in the context of the formation of a semiotic representation, in the sense of Duval.

A second type of graphic operation could be the transformation of a *graphic expression*. This consists of transforming a graphic expression into one, two or more other graphic expressions. The decomposition of a graphic expression can be seen as a particular case of transformation. This type of graphic operation is related to the cognitive activities of processing or conversion. Two types of transformation could be foreseen: those for which the transformed expression denotes the same object, which would be indicated by the fact that the initial and final graphical expressions are equivalent, and those that change the denotation of the graphical expression. We can have several possibilities, depending on whether or not the denotation of the graphic expression is preserved, depending on the treatment or conversion, and also depending on whether, or not, the (theoretical) reference framework is changed, that being taken in the sense of Gonseth (1955) and taken up by Kuzniak (2009) in the definition of a geometric workspace. Table 16.1 summarizes the different transformations.

	Conservation of the denotation		Non-conservation of the denotation	
	Same frame of reference	Different frame of reference	Same frame of reference	Different frame of reference
Treatment				
Conversion				

Table 16.1. *Summary of the different transformations*

Duval (1994, 1995, 2005) has identified three types of graphic transformation associated with operative apprehension: mereological, optical, and positional. Another example of graphic transformation is that of dimensional deconstruction. In the passage from the "surface" vision to the "line" vision or to the "point" vision, certain equivalent graphic expressions do not show the same relations.

The case here is of a transformation that preserves the denotation with the same (theoretical) frame of reference; however, the "donation mode" of the object changes. In this type of operation, we can also be interested in the initial state, the final state or the transformation itself. Rules are necessary to be able to transform a graphical expression, rules that will not be the same depending on what we are doing it for. The work on plot geometry in this book (Chapters 1, 2 and 11) shows various examples, with different constraints in the case of figure restoration (instruments, roughness, cost). But the transformation of graphic expressions is not limited to figures or drawings. Let us take another example with the plan of the classroom. We can imagine transforming this initial plan into another one with a different scale. The denotation of the transformed plan is still the same physical space (the classroom), but the meaning is no longer the same because the transformed plan can show other relationships. It is the same with maps that represent the same physical space but on which different relationships between variables are revealed, depending on the scale. The maps (the graphic expressions) are not identical, but they can be considered equivalent if the denotation is kept the same.

A third type of graphical operation could be the *comparison of graphical expressions*. This would consist of asking about the equivalence of two graphical expressions in the same frame of reference. For example, Mathé and Perrin-Glorian (Chapter 1) raise the question as to the equivalence of construction programs: "exploring the question of equivalence between construction programs, finding reasons for the invalidity or validity of messages seems to provide the opportunity for rich activities, which, even beyond 5th grade, could allow us to accompany pupils in a modification of the object of study, and to engage them in proof checking procedures." Admittedly, this example is not the most appropriate since the construction program is a discursive expression of a procedure for constructing a figure; nevertheless, it allows us to illustrate the question posed. Indeed, the cognitive activity is that of converting one register into another, although the question is indeed that of the equivalence of two construction programs: conservation of the denotation (same geometrical object) by comparing the graphic expressions obtained (the figures). If we obtain the same figure (superimposable or similar), the construction programs are equivalent, even if the drawing does not necessarily follow the same steps or the same sequence.

An example that compares two graphical expressions can be taken from Assude and Gélis (2002). Two figures are constructed with a Dynamic Geometry software; one was constructed as a rhombus and the other as a rectangle. When we open the two files, the graphical expression of the two figures is that of a square, but this is no longer the case when we move them.

In this case, the graphical expressions are not equivalent because their denotation is not the same (in one case, the expression denotes a rectangle, in the other, a rhombus), despite the fact that their initial graphical expression is identical (square).

The graphic space can be an object of study in and of itself, while being associated with that which the graphic expressions can express. The operations envisaged on the figures in Duval's work can be seen from the cognitive point of view as well as from the point of view of graphic operations: hence, we have envisaged naming them differently, while associating them with the cognitive operations of formation, processing and conversion. The implications of this appear to us to be deepening these graphic operations, not only in relation to geometric figures or drawings but also in relation to other graphic expressions essential to geometric work: diagrams, freehand drawings, various coding, etc.

16.3. References

Arzarello, F. (2006). Semiosis as a multimodal process. *Revista Latinoamericana de Investigación en Matemática Educativa*, Special issue on Semiotics, Culture, and Mathematical Thinking, 267–299.

Assude, T. and Gélis, J.M. (2002). Dialectique ancien-nouveau dans l'intégration de Cabri-géomètre à l'école primaire. *Educational Studies in Mathematics*, 50, 259–287.

Barrier, T., Hache, C., Mathé, A.-C. (2014). Droites perpendiculaires au CM2 : restauration de figure et activité des élèves. *Grand N*, 93, 13–37.

Bartolini-Bussi, M. and Mariotti, M.A. (2008). Semiotic mediation in the mathematics classroom: Artefacts and signs after a Vygotskian perspective. In *Handbook of International Research in Mathematics Education*, English, L. (ed.), 2nd edition. Routledge, Taylor and Francis, New York.

Berthelot, R. and Salin, M.-H. (1992). L'enseignement de l'espace et de la géométrie dans la scolarité obligatoire. PhD Thesis, Université Bordeaux 1, Bordeaux.

Bloch, I. and Pressiat, A. (2009). L'enseignement de la géométrie, de l'école au début du collège : situations et connaissances. In *Nouvelles perspectives en didactique des mathématiques*, Bloch, I. and Conne, F. (eds). La Pensée Sauvage, Grenoble.

Bulf, C., Mathé, A.-C., Mithalal, J. (2014). Apprendre en géométrie, entre adaptation et acculturation. Langage et activité géométrique. *Spirale – Revue de recherches en éducation*, 54, 29–48.

Bulf, C., Mathé, A.-C., Mithalal, J. (2015). Langage et construction de connaissances dans une situation de résolution de problèmes en géométrie. *Recherches en didactique des mathématiques*, 35(1), 7–36.

Celi, V. and Perrin-Glorian, M.-J. (2014). Articulation entre langage et traitement des figures dans la résolution d'un problème de construction en géométrie. *Spirale – Revue de recherches en éducation*, 54(1), 151–174.

Chesnais, A. (2018). Un point de vue de didactique des mathématiques sur les inégalités scolaires et le rôle du langage dans l'apprentissage et l'enseignement. Education. Synthesis note, Université de Montpellier.

Chevallard, Y. (1994). Les outils sémiotiques du travail mathématique. *Revue SKOLE*, 1, 51–81.

Drouhard, J.-P. (1995). Algèbre, calcul symbolique et didactique. In *Actes de la 8ème école d'été de la didactique des mathématiques*, Noirfalise, R. and Perrin-Glorian, M.-J. (eds). La Pensée Sauvage, Grenoble.

Duval, R. (1994). Les différents fonctionnements d'une figure dans une démarche géométrique. *Repères IREM*, 17, 121–138.

Duval, R. (1995). *Sémiosis et pensée humaine*. Peter Lang, Bern.

Duval, R. (2005). Les conditions cognitives de l'apprentissage de la géométrie : développement de la visualisation, différenciation des raisonnements et coordination de leur fonctionnement. *Annales de didactique et de sciences cognitives*, 10, 5–53.

Duval, R. (2006). Transformations de représentations sémiotiques et démarches de pensée en mathématiques. *Actes du 32ème Colloque de la COPIRELEM*. IREM, Strasbourg, 67–89.

Duval, R., Godin, M., Perrin-Glorian, M.-J. (2004). Reproduction de figures à l'école élémentaire. In *Actes du séminaire ARDM 2004 de didactique des mathématiques*, Castela, C. and Houdement, C. (eds). IREM Paris 7, 7–89.

Frege, G. (1892). *Sens et dénotation. Écrits logiques et philosophiques*. Le Seuil, Paris.

Gonseth, F (1955). *La géométrie et le problème de l'espace.* Éditions du Griffon, Neuchâtel.

Goody, J. (1979). *La raison graphique.* Les éditions de Minuit, Paris.

Houdement, C. and Kuzniak, A. (2006). Paradigmes géométriques et enseignement de la géométrie. *Annales de didactique et de sciences cognitives,* 11, 175–193.

Kuzniak, A. (2009). Sur la nature du travail géométrique dans le cadre de la scolarité obligatoire. In *Nouvelles perspectives en didactique des mathématiques. Géométrie, les documents pour l'enseignement, le métier de chercheur en didactique,* Bloch, I. and Conne, F (eds). La Pensée Sauvage, Grenoble.

Mathé, A.-C. and Mithalal-Le Doze, J. (2019). L'usage des dessins et le rôle du langage en géométrie: Quelques enjeux pour l'enseignement. In *Nouvelles perspectives en didactique : géométrie, évaluation des apprentissages mathématiques,* Coppé, S., Roditi, E., Celi, V., Chellougui, F., Tempier, F., Allard, C., Corriveau, C., Haspekian, M., Masselot, P., Rousse, S. et al. (eds). La Pensée Sauvage, Grenoble.

Mathé, A.C., Barrier, T., Perrin-Glorian, M.J. (2020). *Enseigner la géométrie à l'école.* Collection Les Sciences de l'éducation aujourd'hui, Academia L'Harmattan, Paris.

Millon-Fauré, K., Roubaud, M.-N., Assude, T. (2019). Entrer dans un genre spécifique : l'écriture d'un programme de construction en géométrie. *Annales de didactique et de sciences cognitives,* 24, 9–45.

Perrin-Glorian, M.-J. and Godin, M. (2018). Géométrie plane : pour une approche cohérente du début de l'école à la fin du collège [Online]. Available at: https://hal.archives-ouvertes.fr/hal-01660837v2/document.

Radford, L. (2014). On the role of representations and artefacts in knowing and learning. *Educational Studies in Mathematics,* 85, 405–422.

17

Towards New Questions in Geometry Didactics

This chapter takes contributions from the book and puts them in dialogue with other research in geometry didactics.

This work leads to recommendations for the pursuit of certain questions and uncovers new ones.

17.1. Current questions in geometry didactics

In 2003, during a conference aimed at establishing state of the art didactic research, Perrin-Glorian identified five major questions which, according to her at that time, arose in geometry didactics (Perrin-Glorian 2004). We take up these questions here and relate them to different chapters within our book.

What content to teach? What are the objectives of geometry teaching? More broadly, what do we call geometry in elementary school? It appears that these questions remain alive and well in 2022, whatever the age of the pupils (see Chapters 1, 3–5, 9 and 14). Moreover, they are addressed in this book in connection with the other lines of questioning.

Which learning difficulties? This research theme seems to have evolved. Indeed, particular attention has been paid to difficulties of a non-mathematical nature, notably those linked to praxis and visuo-spatial deficiencies (see the work of Petitfour 2015). On the other hand, several recent works focus instead on the consideration of difficulties (mathematical or otherwise), or even question the ways

---

Chapter written by Claire GUILLE-BIEL WINDER and Catherine HOUDEMENT.

in which they are taken into account (see Chapters 2, 8, 9, 13). The difficulties relating to the question of "relations between the theoretical and material or visual aspects of geometric objects" (Perrin-Glorian 2004, p. 42, *translated by author*) lead us to reflect on the link between "graphical space" and "geometric space". In addition, the interest shown by several researchers into this work, concerning graphical objects, appears crucial to us. It also concerns the following point.

Which learning situations? This question is always at the heart of studies, as can be seen from the numerous chapters that deal with it, whether it concerns spatial knowledge or geometric knowledge (Chapters 1–3, 6, 7, 9–11, 14). In this book, it is coupled with a reflection on the resources made available or to be made available to teachers, and even on the conditions of this availability, which is in line with the following questioning.

Which teacher training? Which classroom practices? These two questions, which were under-addressed in 2003 (Perrin-Glorian 2004), are now the subject of more works (Chapters 1, 3, 9–15). These are sometimes related to each other (training–practice relationship) or even to classroom resources.

Moreover, the question of the place and role of language in geometry, not raised as such in 2003, now seems to us as crucial: it is currently the subject of many works (let us cite, for example, Mathé 2012; Celi and Perrin-Glorian 2014; Barrier et al. 2014; Bulf et al. 2015; Chesnais 2018; Mathé and Mithalal 2019; Guille-Biel Winder 2021; or Mathé et al. 2021), and is a theme that runs through most of the chapters in this book. Several chapters study and refine activities based on the spatial apprehension of objects (Chapter 3) and figures, developing a particular interest in actions traced with variable instruments (Chapters 1, 2, 5, 6, 8, 11, 12). Several aims intersect: enriching spatial apprehension through "geometry", by planning to have the pupils work on the visualization of figures and/or on language description. The collective project is clear: to propose rather new activities, quite motivating for the pupils and potentially carrying geometrical knowledge. It could be described as a reflection on the balance between teaching work on drawing and work on language, the instrumented action of pupils, with language accompanying drawing decisions, and the descriptions of drawings, first spatial, then increasingly geometric. This project is also designed to train teachers in a new, more serene relationship with school plane geometry, with the need to teach alternative visualizations of a figure, and a motivation of geometric language through work on instrumented drawing. How can we then speak of continuity between elementary school geometry and middle school geometry? How can teachers be acculturated to this new approach to school geometry?

These two themes are developed in this chapter. The first (section 17.2) studies the continuities and ruptures between "graphical space" and "geometrical space" that have been uncovered throughout the book. The second (section 17.3) addresses the question on the articulation between resources for teaching and learning geometry, teaching practices and training "devices" (or modalities).

17.2. Continuities and breaks in the teaching of geometry

In Chapter 1 (Perrin-Glorian and Mathé), "possible continuities between material geometry and theoretical geometry" are evoked, reformulated elsewhere as links between geometric drawing, or even the "geometry of tracing" and "theoretical geometry". The question of continuities between primary school and middle school is not new. In this section, we study continuities at different scales.

17.2.1. *Institutional continuity?*

With regard to institutions and their functioning (Chevallard 1992), there are many visible breaks between primary school and middle school that impact teaching: physically different places of reception; unity of place (the classroom) for primary school, but not for middle school; type of teacher: generalist or specialist in one or two subjects; one/two teachers per class group or many who follow one another during the day; variable and fairly flexible duration of the session at school, set at 55 minutes at middle school. Overall, continuity thus seems compromised, etc.

Since the *Loi d'Orientation* of 2013, the primary school–middle school curricula have been brought together. In France, for Cycle 3 (of middle school pupils aged 9–12 years old), there is a single document for Mathematics (effective since the start of the 2015 school year) with indications by class level. This shows an institutional project of continuity between primary school and middle school in the teaching injunctions.

17.2.2. *Theoretical continuity from "geometry of tracing" to "abstract geometry"?*

Following the anthropological theory of didactics (Chevallard 1999), we consider the mathematical activity in an institution organized in two blocks: the theoretical block (technological–theoretical, knowledge) and the praxis block (tasks and techniques). We question the thread of Chapter 1 in the quotation "[the geometry of tracings] corresponds to the search for a possible link in learning between tracing lines with instruments and the notions of abstract geometry" (Chapter 1, p. 8) in light of this praxeological organization.

In order to describe the different geometries called upon by institutions and/or put into activity by teachers, Houdement and Kuzniak use the term "paradigms" (in the sense of Kuhn (1983)), which embodies the set of beliefs and techniques shared by a scientific group, the appropriate ways of solving a problem, and even the significant examples proposed to learners to acculturate them to the paradigm (Houdement and Kuzniak 1999, 2006; Houdement 2007). Two paradigms, epistemologically founded, are thus called upon in the compulsory teaching of elementary geometry: Geometry 1 and Geometry 2. More exactly, these two paradigms represent two "horizons" of geometric work, especially in school, as we will see later.

Geometry 1 takes as its objects of study material objects: graphical traces on a sheet of paper or virtual traces on a screen; preliminary models of real objects or situations. Thought is exercised on these objects through immediate or constructed perception, experiments (tracing, folding, cutting, etc.), as well as reasoning that triggers and takes advantage of these experiments. Geometry 2 takes as objects of study "ideal objects" (Houdement 2007), defined in a textual way, from elementary forms used to describe and measure space. It organizes (orders), by hypothetico-deductive laws, the results of Geometry 1, based on certain results considered as acquired (the axiomatic base) which it extends and enriches. Geometry 1 is an experimental ground for Geometry 2; it integrates the use of material instruments.

Geometry 2 is a theoretical model of Geometry 1; there is no material instrument in Geometry 2, only an intellectual instrument: the hypothetico-deductive reasoning which takes place through texts and figures for a better understanding of the argument. This fundamental difference in the modes of production of new results between Geometry 1 and Geometry 2 is not an intrinsically geometrical necessity according to Szabó (2000). The refutation of visual verification, which is echoed in Euclid's "Elements", a treatise in essence of Geometry 2, would be born of the appearance of a new type of proof, the *reductio ad absurdum* (used to prove the irrationality of $\sqrt{2}$, "the square root of 2"), which can only be exercised on ideal objects, in contrast to the synthetic demonstration that could be based on the real (Houdement 2007, p. 71). Epistemologically, there is thus a break between Geometry 1 and Geometry 2.

However, for teaching, there are possible situations to switch from Geometry 1 to Geometry 2 (Berthelot and Salin 2001; Houdement and Rouquès 2016; Houdement 2019). The most famous one in didactics is the "fundamental situation of the study of geometry" (in the sense of Brousseau's Geometry 2 [1983, 2001]). In this situation, the teacher presents pupils with a fake drawing: the teacher shows a sheet of paper on which they have drawn a triangle and its perpendicular bisectors. However, these perpendicular bisectors do not intersect at a single point as perpendicular bisectors normally do in a triangle: in the teacher's drawing, they form a small triangle, which the teacher calls the "friend triangle". The pupils' task is to construct a triangle that "has" a bigger friend triangle. The pupils' activity is thus an instrumented construction. However, the impossibility of constructing a bigger friend triangle, whatever the starting triangle, leads the pupils to protest, or even declare the teacher's question impossible. The teacher then changes the nature of the task: no longer tracing, but questioning the enlargement task. They engage (and help) the pupils to call upon (geometric) language arguments, linked to the relations between points of a segment and points of the perpendicular bisector of the segment (which, incidentally, deconstructs the figure, from the lines to the points). The collective argumentation (pupils AND teacher) concludes that it is necessary (mathematically) for the perpendicular bisectors to meet at a single point, whatever the starting triangle. It reveals the "theoretical illusion" of the friend triangle drawn. The predominance of linguistic arguments that allow for deducing from geometrical knowledge already there – any point of the perpendicular bisector of a segment is equidistant from the extremities of the segment – the necessity of the property: the three perpendicular bisectors intersect at a single point – marks the entry into Geometry 2. This provides an explanation for the impossibility of enlarging the friend triangle and leads to a differentiation between the "perceived" and the "known" (Parzysz 1988). It is moreover significant to give a name to this new "geometric practice" (e.g. Geometry 2) in front of the pupils, to differentiate it from the previous geometry (Houdement and Rouquès 2016). It is the occasion to declare, to ratify the irrelevance, in Geometry 2, of a (visual) control even equipped with the instruments (standard geometric instruments or not) or material actions of comparing between figures (superposition, composition, decomposition, etc.). This part of the fundamental situation of geometry comes under the acculturation at the technologic-theoretical level of the punctual praxeology (Chevallard 1999) that represents the situation of construction of the friend triangle of a triangle (which is not exactly the Brousseau situation). The technological discourse invalidates the existence of a friend triangle, and produces and validates the concurrence of the perpendicular bisectors of a triangle, whatever the triangle is. Entering Geometry 2 requires access to a technological discourse on Geometry 1, but not just any discourse: one whose function is the validation (Castela and Elguero 2013) of an action in Geometry 1.

However, the "difference" between Geometry 1 and Geometry 2 cannot be reduced to a change of relationship to the figure. The nature of the geometric objects at stake in each paradigm, the working techniques, are profoundly different. The two paradigms coexist in the "geometric workspace", a notion introduced by Houdement and Kuzniak (2006), to account for these two facets of geometric work. Both the expert and the novice take advantage of the potentiality of Geometry 1 (especially of the effective traces), but it is the ability to consider the results coming from Geometry 1 as conjectures, and to retain only the results coming from Geometry 2 as characterizing the expert from the novice (Houdement 2007). The shift to theoretical geometry corresponds to a paradigmatic shift in the horizon for geometric work.

17.2.3. *Praxis continuity from the "geometry of tracing" to "abstract geometry"*

Throughout this book, there are reflections on the "geometry of tracing", defined in the first chapter and studied in the following chapters. Let us clear up a persistent misunderstanding: Geometry 1 does not contain any restrictions on the instruments to be used, nor on the actions to be taken in order to produce new figures or to validate (folding, cutting, superimposing); the validation of the correctness of figures is pragmatic:

> The tasks can be specified by the choice of authorized instruments: for example, measuring is a legal and common technique in Geometry 1, but there are also problems that can be solved in Geometry 1 without measuring (Duval 2005; Keskessa, Perrin and Delplace 2007). The usual experiment in this paradigm is the "instrumented drawing". (Houdement 2007, p. 73, *translated by author*).

17.2.3.1. *The restoration of figures*

The "geometry of tracing" excludes measurement and focuses on geometrical quantities (lengths, angles). This is an important change of geometric perspective for teachers, for whom measurement often takes precedence over form, notably due to the use of standard instruments (the graduated ruler and the square, also sometimes graduated) and perhaps because of the etymology of the word "geometry", as "the measurement of the earth".

In the figure reproduction situations developed in this book, the instruments are variables of the instrumented actions, and the validation of the action is done by superimposing the obtained figure with a solution layer. The instruments are chosen to give the possibility to formulate a geometrical property after having "acted" upon it: for example, two points are (always) aligned, a segment supports a single line, two lines (not parallel) intersect in a single point, etc. This linguistic formulation of properties, under the teacher's supervision, makes it possible to use geometric language, and even to write it. In this sense, and thanks to the action of the teacher who triggers the verbalization of the pupils and reformulates them in geometric language, the "geometry of tracing" prepares for the adapted use of geometric language, in particular for the acculturation to the linguistic formulation of geometric properties. But we know that this is not so simple in practice: Blanquart (2020) shows the fundamental role of the teacher in offering pupils the possibility of access to geometric language, and not only to the geometric vocabulary to which it is often reduced, especially by school programs. In the same spirit, Guille-Biel Winder (2021) compares the richness of the geometric vocabulary of young pupils in two classes and reveals the correlation with the language offered by each teacher.

Abstract or theoretical geometry (according to the authors' choice) corresponds to Geometry 2. Geometry 2 intrinsically mobilizes two semiotic systems: the figure, of course, as well as the text, and in particular the written text. A geometric object (of Geometry 2), or even a geometric problem, can be completely defined (1) by a text, (2) by a text accompanied by a figure, (3) by a coded figure. The cases (2) and (3) are close because the codes are conventional graphical signs placed on the figure which save the explicit writing of geometrical properties, such as the equality of lengths, angles, perpendicularity. It is often the case that the text alone does not give an account of the geometric object: the properties of incidence (point on, and intersection of lines) and of alignment, for example, are often not made explicit, but are nonetheless visible on the figure. In any case, text and figure are rarely redundant.

It is true that "the objects of theoretical geometry are related by statements that translate into visual or instrumental features of physical geometric figures" (Chapter 1, p. 4), but in a geometric demonstration problem, not all "visual or instrumental features of physical figures" are theoretical statements (Laborde 1988, *translated by author*). This non-reciprocity is one of the characteristics for the change in relation to figures between Geometry 1 and Geometry 2: many works (as early as Laborde 1988) have shown that taking it into account is very complex for middle school pupils dealing with these geometric problems.

Therefore, the "geometry of tracing" does not intrinsically involve a relationship to the geometric text, or even to language, as the authors specify: "The reproduction of a figure by someone who has both the starting figure and the model does not require the use of language" (Chapter 1, p. 13); it is a question of reproducing figures, and therefore of processing them within the register of figures (Duval 1995, 2005).

The figures under study in the "geometry of tracing" always seem to be assemblages of basic figures, in other words complex figures. Restoration activities, like Duval (2005) and Duval and Godin (2005), teach the pupil to mobilize different views of the figure under study; they enrich the possible visualizations of a figure, lead to consider it as an assembly of surfaces, a network of lines or a network of points. These decompositions/re-compositions are in fact a necessary condition for identifying the constituent properties of the figure (alignments of points and segments, relative positions of two lines, equalities of lengths or even angles), and thus reproducing it. The "geometry of tracing" enriches the work in Geometry 1 and contributes to the foundation of the work covered in Geometry 2. In the rest of the book, it is predominantly the tracing activities that are studied.

17.2.3.2. *Working on language*

In the second part of Chapter 1 (Perrin-Glorian and Mathé), the authors work more specifically on language, as based on Brousseau's typology of didactic situations, with situations of formulation and validation. This completely changes the meaning ascribed to "geometry of tracing":

> The definition of "Geometry of tracing" aims to identify a field of practices and discourses that can be held on physical geometric figures to facilitate the entry into theoretical geometry (Chapter 1, p. 9).

The task is still to *produce* a figure, in a situation where it is not visible to the *constructor* (who does the drawing), and it is up to the *instructor* (who has the model figure) to write a message for the constructor to produce the expected figure. We focus on the message texts, as the authors do. The text is thus a communication tool; hence, it is the conformity a priori of the figure produced that should validate success. How, then, can we ensure that the message produced is written in geometric language, which for all intents and purposes appears to be an aim of the activity? How can pupils acquire this language? The examples (such as Figure 1.8) show the need for pupils to have been confronted, before these formulation sessions, with construction tasks: indeed, these construction tasks could serve as acculturation to geometric texts, and thus to geometric language in a situation. In our view (Houdement and Petitfour, 2022), activities involving the construction of a figure from a written text (e.g. a construction program, but not only) are really necessary for geometric learning, especially in Grades 4–6 (in France known as Cycle 3).

NOTA BENE.– Construction problems, especially with the ruler and the compass (a constraint in terms of the instruments available), have always been constituted as practices of acculturation to geometry (e.g. Petersen 1990). These problems explicitly call upon two semiotic systems, the text and the figure, and require a conversion in the sense of Duval (2005), this conversion leading the geometrician to work with a horizon, sometimes Geometry 1, sometimes Geometry 2, in order to mobilize and deduce the properties necessary for the construction.

Later on, the authors present examples of such "construction tasks" (Chapter 1, Figure 1.12) as examples of valid messages, and then work on the equivalence between "construction tasks" (section 1.6.1). In this part, we also understand why it is important that the restoration situations have been accompanied by a collegial formulation (teacher and pupils), which the authors have termed as "practical axiomatics".

In brief, this book renews geometry situations, especially the reproduction of figures, mediated by instruments chosen to articulate specific properties in words, especially those that serve as a foundation for future theoretical geometry. This dimension of research seems to us relatively stable and operational, insofar as the proposed situations offer a certain a priori reproducibility. Several chapters of this book offer perspectives for working on language, based on figures (or objects), and in particular their description. This dimension, which is in itself complex, deserves to be developed in greater detail in order to create situations whose effects can be anticipated and controlled by the teacher.

17.3. Articulation between resources, practices and teacher training

The question of the articulation between resources, practices and teacher training is addressed more or less explicitly in several chapters of this book. It is often correlated with the use of research that results in the development of resources (Chapters 1, 6, 9, 11, 14) or within systems aimed at changing practices (Chapters 3, 6, 7, 11, 14, 15). In our opinion, this expands in several directions.

What situations should be used to develop the pupils' relationship with figures? What consideration should be given to spatial knowledge? What professional gestures should teachers implement? Assude (Chapter 6) presents a set of 10 situations for the Grade 1 and Grade 2 pupils (aged 7–9 years), alternating paper–pencil activities, or in meso-space, with an activity using dynamic geometry software. Concerning this same age group, the ERMEL team (Chapter 9) proposes didactic situations in progression, for the learning of straight lines, figures and solids, aiming at questioning the erroneous conceptions of the pupils (in particular those related to perception), and at making emerge new criteria of judgment generated from the acquired geometrical knowledge. Marchand and Bisson (Chapter 14) have developed teaching (in primary schools) and training sequences, dealing with the reproduction of figures and using a particular artifact (Tangram), with the aim of developing spatial knowledge. The development of figure restoration situations (Perrin-Glorian and Mathé; Chapter 1; Vièque, Chapter 2; Mangiante-Orsola, Chapter 11), as well as their evolution towards formulation and proof situations (Chapter 1) testify to the interest in teaching students on how to look at a figure. As these situations have been elaborated and tested within the framework of research, the question of their implementation to standard teaching practices arises (Chapter 11). In research on the reproduction of plane figures at the end of primary school and the beginning of secondary school (pupils aged 11–13 years), Blanquart (Chapter 13) underlines, indeed, the difficulty facing teachers who rely on pupils' reasoning during the process of institutionalization, and highlights the importance of a preliminary analysis for a better observation of the pupils' processes. Finally, she hypothesizes that a lack of didactic knowledge (Houdement 2013) is at the root of these difficulties. This work thus joins the following broader questioning: Under what conditions can teachers exercise didactic vigilance (Charles-Pézard 2010; Guille-Biel Winder and Mangiante-Orsola 2023) when teaching geometry?

What impact can the use of resources have on teaching and learning geometry? Can we change practices through the use of resources? This is the hypothesis explicitly retained by Mangiante-Orsola (Chapter 11) when she developed a resource aimed at provoking change in teacher practices. This hypothesis also underlies the work of the ERMEL team (Chapter 9), which aims, through their resource, to provide teachers (and their trainers) with tools for analysis (of the choices made by the resource, as well as of their own practice), so that they can implement the situations proposed. It is likewise a central tenet of Assude's line of questioning (Chapter 6). It remains to be confirmed.

Can practices be changed through the development of resources? The Celi experiment (Chapter 3) is part of a collaborative research project. However, the enrichment of practices that potentially promote learning is not systematically observed, despite the support of training sessions (Mangiante-Orsola, Chapter 11) or collaborative support (Celi, Chapter 3). This leads to the following broader questioning.

Under what conditions can teaching practices in geometry evolve? Certain necessary conditions have already been identified: the proposed evolutions must be in line with teachers' professional concerns and must not be too destabilizing for them (Butlen, Mangiante-Orsola and Masselot 2017); they must be able to adjust to old practices (Assude and Gélis 2002), and in the case of the use of dynamic geometry software, allow for a certain time saving within the didactic system (Assude 2005). But are these conditions sufficient? Various teacher training modalities, resulting from didactic research, have already been explored, especially those in the numerical domain: homology-transposition situations (Houdement and Kuzniak 1996), didactic analysis of practices, application research strategies (Houdement 2013), and teacher training through coaching (Ngono and Peltier-Barbier 2004).

However, the effects of these teacher training sessions on practice are understudied. Observing five teachers in class after they and other volunteers had taken a 48-hour training course in geometry, Vergnes (2004) concludes that the effects on practice are mixed, in particular because attempts to change practices generate "costly incidents for them in terms of classroom management" (Vergnes 2004, p. 196, *translated by author*).

Moreover, new teacher training modalities have been developed since the early 2000s, whose potentialities are the subject of a number of works, such as role-playing games (Lajoie and Pallascio 2001; Lajoie 2010; Lajoie et al. 2019), collaborative research (Desgagné 1997, 2007; Bednarz 2013) or lesson studies (Miyakawa and Winsløw 2009; Clivaz 2018). However, these studies are not specifically in the geometry domain. What teacher training, in the context of teaching/learning geometry, should be considered (form, content)? Would there be more relevant teacher training modalities to favor this evolution?

Research in geometry didactics has a bright future ahead of it!

17.4. References

Assude, T. (2005). Time management in the work economy of a class. A case study: Integration of Cabri in primary school mathematics teaching. *Educational Studies in Mathematics*, 59(1), 183–203.

Assude, T. and Gélis, J.M. (2002). Dialectique ancien-nouveau dans l'intégration de Cabri- géomètre à l'école primaire. *Educational Studies in Mathematics*, 50, 259–287.

Barrier, T., Hache, C., Mathé, A.-C. (2014). Droites perpendiculaires au CM2 : Restauration de figure et activité des élèves. *Grand N*, 93, 13–37.

Bednarz, N. (2013). *Recherche collaborative et pratique enseignante. Regarder ensemble autrement*. L'Harmattan, Paris.

Berthelot, R. and Salin, M.-H. (2001). L'enseignement de la géométrie au début du collège. Comment concevoir le passage de la géométrie du constat à la géométrie déductive ? *Petit x*, 56, 5–34.

Blanquart, S. (2020). Raisonnements géométriques d'élèves de cycle 3, duos de situations, rôle de l'enseignant. PhD Thesis, Université de Paris [Online]. Available at: https://tel.archives-ouvertes.fr/tel-03242768/ [Accessed 3 March 2022].

Brousseau, G. (1983). Étude des questions d'enseignement. Un exemple ; la géométrie. *Séminaire de didactique des mathématiques et de l'informatique*. IMAG, Université Fourier, Grenoble, 183–226.

Brousseau, G. (2001). Les propriétés didactiques de la géométrie élémentaire. L'étude de l'espace et de la géométrie. *Actes du séminaire de didactique des mathématiques*, 2000. University of Crete, 67–83.

Bulf, C., Mathé, A.-C., Mithalal, J. (2014). Apprendre en géométrie, entre adaptation et acculturation. Langage et activité géométrique. *Spirale – Revue de recherches en éducation*, 54, 29–48.

Butlen, D., Mangiante-Orsola, C., Masselot, P. (2017). Routines et gestes professionnels, un outil pour l'analyse des pratiques effectives et pour la formation des pratiques des professeurs des écoles en mathématiques. *Recherches en didactiques*, 2017(2), 25–40.

Castela, C. and Elguero, C. (2013). Praxéologie et institution, concepts clés pour l'anthropologie épistémologique et la socio-épistémologie. *Recherches en didactique des mathématiques*, 33(2), 123–162.

Celi, V. and Perrin-Glorian, M.-J. (2014). Articulation entre langage et traitement des figures dans la résolution d'un problème de construction en géométrie. *Spirale – Revue de recherches en éducation*, 54(1), 151–174.

Charles-Pezard, M. (2010). Installer la paix scolaire, exercer une vigilance didactique. *Recherches en didactique des mathématiques*, 30(2), 197–261.

Chesnais, A. (2018). Un point de vue de didactique des mathématiques sur les inégalités scolaires et le rôle du langage dans l'apprentissage et l'enseignement. Education. Synthesis note, Université de Montpellier.

Chevallard, Y. (1992). Concepts fondamentaux de la didactique : perspectives apportées par une approche anthropologique. *Recherches en didactique des mathématiques*, 12(1), 73–112.

Chevallard, Y. (1999). L'analyse des pratiques enseignantes en théorie anthropologique du didactique. *Recherches en didactique des mathématiques*, 19(2), 221–266.

Clivaz, S. (2018). Développement des connaissances mathématiques pour l'enseignement au cours d'un processus de lesson study. In *Actes du séminaire national de didactique des mathématiques 2016*, Barrier, T. and Chambris, C. (eds). IREM de Paris, Université Paris Diderot, 287–302.

Desgagné, S. (1997). Le concept de recherche collaborative : l'idée d'un rapprochement entre chercheurs universitaires et praticiens enseignants. *Revue des sciences de l'éducation*, 31(2), 245–258.

Desgagné, S. (2007). Le défi de coproduction de "savoir" en recherche collaborative, analyse d'une démarche de reconstruction et d'analyse de récits de pratique enseignante. In *La recherche participative : multiples regards*, Anadòn, M. and Savoie-Zajc, L. (eds). Presses de l'université du Québec.

Duval, R. (1995). *Sémiosis et pensée humaine*. Peter Lang, Bern.

Duval, R. (2005). Les conditions cognitives de l'apprentissage de la géométrie : développement de la visualisation, différenciation des raisonnements et coordination de leur fonctionnement. *Annales de didactique et de sciences cognitives*, 10, 5–53.

Duval, R. and Godin, M. (2005). Les changements de regard nécessaires sur les figures. *Grand N*, 76, 7–27.

Guille-Biel Winder, C. (2021). Impact du langage de l'enseignant sur les relations entre les élèves et le milieu dans une situation d'action en géométrie. *Recherches en didactique des mathématiques*, 41(1), 55–96.

Guille-Biel Winder, C. and Mangiante-Orsola, C. (2023). Contribution à l'étude de l'exercice de la vigilance didactique. *Recherches en didactique des mathématiques*.

Houdement, C. (2007). À la recherche d'une cohérence entre géométrie de l'école et géométrie du collège. *Repères-IREM*, 67, 69–84.

Houdement, C. (2013). Au milieu du gué : entre formation des enseignants et recherche en didactique des mathématiques. Synthesis note, Université Paris Diderot, Université de Rouen. Available at: https://hal.archives-ouvertes.fr/hal-03201021/ [Accessed 3 March 2022].

Houdement, C. (2019). Spatial et géométrique, le yin et le yang du géométrique. In *Nouvelles perspectives en didactique : géométrie, évaluation des apprentissages mathématiques*, Coppé, S., Roditi, E., Celi, V., Tempier, F., Allard, C., Corriveau C., Haspekian, M., Masselot, M., Rousse, S., Sabra, H. (eds). La Pensée Sauvage, Grenoble.

Houdement, C. and Kuzniak, A. (1996). Autour des stratégies utilisées pour former les maitres du premier degré en mathématiques. *Recherches en didactique des mathématiques*, 16(3), 289–322.

Houdement, C. and Kuzniak, A. (1999). Sur un cadre conceptuel inspiré de Gonseth et destiné à étudier l'enseignement de la géométrie en formation des maîtres. *Educational Studies in Mathematics*, 40(3), 283–312.

Houdement, C. and Kuzniak, A. (2006). Paradigmes géométriques et enseignement de la géométrie. *Annales de didactiques et de sciences cognitives*, 11, 175–193

Houdement, C. and Petitfour, E. (2022). Le dessin à main levée, un révélateur du rapport des élèves à la figure géométrique. *Canadian Journal of Science, Mathematics and Technology*, 22, 315–340.

Houdement, C. and Rouquès, J.P. (2016). Deux géométries en jeu dans la géométrie plane. *Contribution aux ressources Éduscol sur la géométrie plane* [Online]. Available at: https://hal.archives-ouvertes.fr/hal-03214099v2/document.

Kuhn, T.S. (1983). *The Structure of Scientific Revolutions*. French translation: *La structure des révolutions scientifiques*. Flammarion, Paris.

Laborde, C. (1988). L'enseignement de la géométrie en tant que terrain d'exploitation de phénomènes didactiques. *Recherches en didactique des mathématiques*, 9(3), 337–364.

Lajoie, C. (2010). Les jeux de rôles : une place de choix dans la formation des maîtres du primaire en mathématiques à l'UQAM. In *Formation des enseignants en mathématiques : tendances et perspectives actuelles*, Proulx, J. and Gattuso, L. (eds). Éditions du CRP, Sherbrooke

Lajoie, C. and Pallascio, R. (2001). Role-play by pre-service elementary teachers as a means to develop professional competencies in teaching mathematics. *Proceedings of SEMT'01 – International Symposium Elementary Mathematics Teaching*. Charles University, Prague.

Lajoie, C., Mangiante, C., Masselot, P., Tempier, F., Winder Guille-Biel, C. (2019). Former à aider un élève en mathématiques. Une étude des potentialités d'un scénario de formation basé sur un jeu de rôles. *Revue canadienne de l'enseignement des sciences, des mathématiques et de la technologie/Canadian Journal of Science, Mathematics and Technology Education*, 19, 168–188.

Mathé, A.-C. (2012). Jeux et enjeux de langage dans la construction de références partagées en géométrie. *Recherches en didactique des mathématiques*, 32(2), 195–227.

Mathé, A.-C. and Mithalal-Le Doze, J. (2019). L'usage des dessins et le rôle du langage en géométrie : quelques enjeux pour l'enseignement. In *Nouvelles perspectives en didactique : géométrie, évaluation des apprentissages mathématiques*, Coppé, S., Roditi, E., Celi, V., Tempier, F., Allard, C., Corriveau C., Haspekian, M., Masselot, M., Rousse, S., Sabra, H. (eds). La Pensée Sauvage, Grenoble.

Mathé, A.-C., Maillot, V., Ribennes, J. (2021). Enjeux langagiers, situations de formulation et de validation en géométrie. Un exemple de travail autour du cercle en CE2. *Grand N*, 108, 27–57.

Miyakawa, T. and Winsløw, C. (2009). Un dispositif japonais pour le travail en équipe d'enseignants : étude collective d'une leçon. *Education et didactique*, 3(1), 77–90.

Ngono, B. and Peltier-Barbier, M.-L. (2004). Les effets d'une formation par acompagnement sur le temps long sur les pratiques de professeurs confirmés. In *Dur d'enseigner en ZEP*, Peltier-Barbier, M.-L. (ed.). La Pensée Sauvage, Grenoble.

Parzysz, B. (1988). "Knowing" vs "seeing". Problems of the plane representation of space geometry figures. *Educational Studies in Mathematics*, 19(1), 78–91.

Perrin-Glorian, M.-J. (2004). Vingt ans de didactique en 1993 ! Où en est-on dix ans après ? In *Actes du 30ème colloque inter-IREM des professeurs et formateurs de mathématiques chargés de la formation des maîtres (Avignon, 19-21 May 2003)*. IREM de Marseille, 33–78.

Perrin-Glorian, M.-J. and Godin, M. (2018). Géométrie plane : pour une approche cohérente du début de l'école à la fin du collège [Online]. Available at: https://hal.archives-ouvertes.fr/hal-01660837v2/document.

Petersen, J. (1990). *Problèmes de constructions géométriques*. Éditions Jacques Gabay, Sceaux.

Szabó, Á. (2000). *L'aube des mathématiques grecques*. Paris, Vrin.

Petitfour, E. (2015). Enseignement de la géométrie à des élèves en difficulté d'apprentissage : étude du processus d'accès à la géométrie d'élèves dyspraxiques visuo-spatiaux lors de la transition CM2-6ème. PhD Thesis, Université Paris Diderot.

Vergnes, D. (2004). Effets d'un stage de formation en géométrie sur les pratiques d'enseignants de l'école primaire. In *Dur d'enseigner en ZEP*, Peltier-Barbier, M.-L. (ed.). La Pensée Sauvage, Grenoble.

Appendices

Appendix 1

Four Situations

Situation 1: blue and red circles

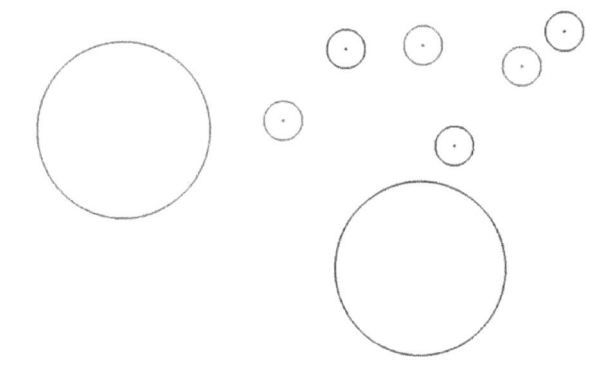

Figure A1.1. *Figure given at the start of situation 1. For a color version of this figure, see www.iste.co.uk/guille/tangible.zip*

Quick description	The pupils must observe the figure, describe it and then move the small red circles inside the large red circle and the small blue circles inside the large blue circle. Then they must use the track for these movements and observe the trajectories.
Types of tasks covered	Ti1: Move points and objects. Ti2: Use the trace to visualize trajectories. Tg1: Describe a simple figure. Tg2: Use the appropriate geometric vocabulary (circle). Ts1: Identify certain spatial relationships (interior, exterior).

Table A1.1. *Situation 1*

Situation 2: little ants

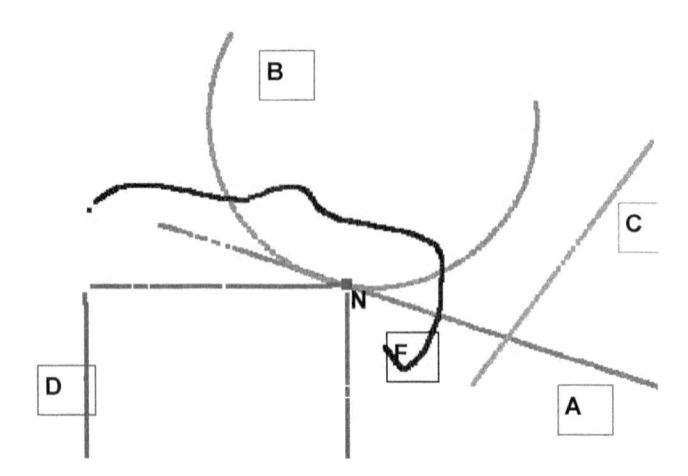

Figure A1.2. *Type of figure targeted in situation 2*

Quick description	The pupils must move points which can move everywhere, points which move on trajectories defined a priori and observe that there are points which cannot move directly. Then, the students must use the trace to recognize the circle, the square and the segment as the trajectory of points.
Types of tasks covered	Ti1: Move points and objects. Ti2: Use the trace to visualize trajectories. Ti3: Recognize the different "cabri" points (free point, constrained point and dependent point). Tg2: Use the appropriate geometric vocabulary (circle, square, segment). Tg3: Recognize geometric figures as point trajectories (segments, circles, square).

Table A1.2. *Situation 2*

Appendix 1 307

Situation 3: doll

Figure A1.3. *Figure given at the start of situation 3*

Quick description	The students must observe the doll and identify the different elements that make it up. They must then move these different elements and then create a doll by building segments, circles and triangles.
Types of tasks covered	Ti1: Move points and objects. Ti4: Use menus to trace objects. Tg4: Describe a complex figure by recognizing simple geometric figures (segment, circle, triangle). Rg5: Trace segments, circles and triangles.

Table A1.3. *Situation 3*

Situation 4: compositions

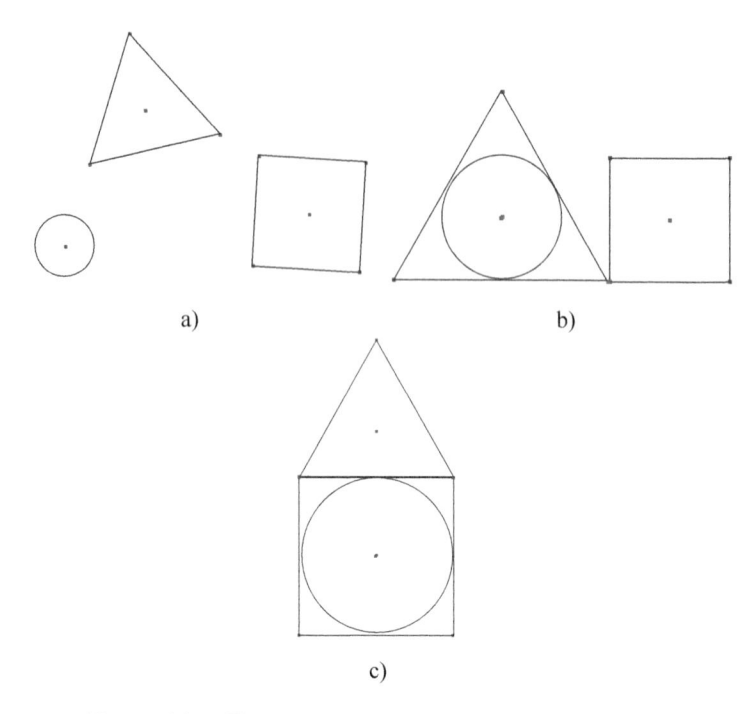

Figure A1.4. *Figure to observe and reproduce in the first stage of situation 4: a) starting file (Cabri); b) and c) two figures to be reproduced (data on paper and pencil)*

Quick description	The pupils must reproduce a given complex figure with pencil and paper using a Cabri file that already contains the three basic figures (a triangle, a square and a circle).
	In a second step, in Cabri, the pupils create a complex figure using the three basic figures and then they must reproduce it on paper using a pencil.
Types of tasks covered	Ti1: Move points and objects.
	Tg4: Describe a complex figure by recognizing simple geometric figures.
	Rg6: Reproduce a complex figure from simple figures.

Table A1.4. *Composition of figures*

Appendix 2

FOLDING AND SYMMETRY Situation

The description below (FOLDING AND SYMMETRY Situation, Theme 2) includes:

– a link to download the digital documents needed for the situation;

– a summary of the problem posed, the initial and targeted knowledge, and the possible organization in class;

– a detailed description, for each phase, of the activities, the instructions, the functions (presentation, research, sharing) and the forms of work (first column);

– didactic and pedagogical comments presenting the productions observed and the possible choices of the teacher (second column).

General information (problem posed, knowledge involved, etc.) is given at the outset.

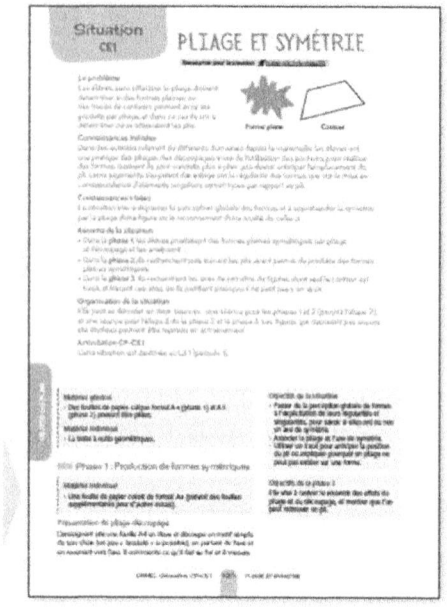

The description of the procedure (material, instructions, etc.) allows, from its first reading, a reliable implementation.

The specific issues and the organization of each phase in detail.

Appendix 3

Support Systems

Ready-to-make material: a materials kit is available. It contains all the pre-cut patterns and shapes used in the situations. Teachers have all the solid patterns available for download in pdf format at https://www.hatier-clic.fr/2512565. Files for 3D printing are also available. Teachers can, for example, approach a fablab to print or print the solids themselves.

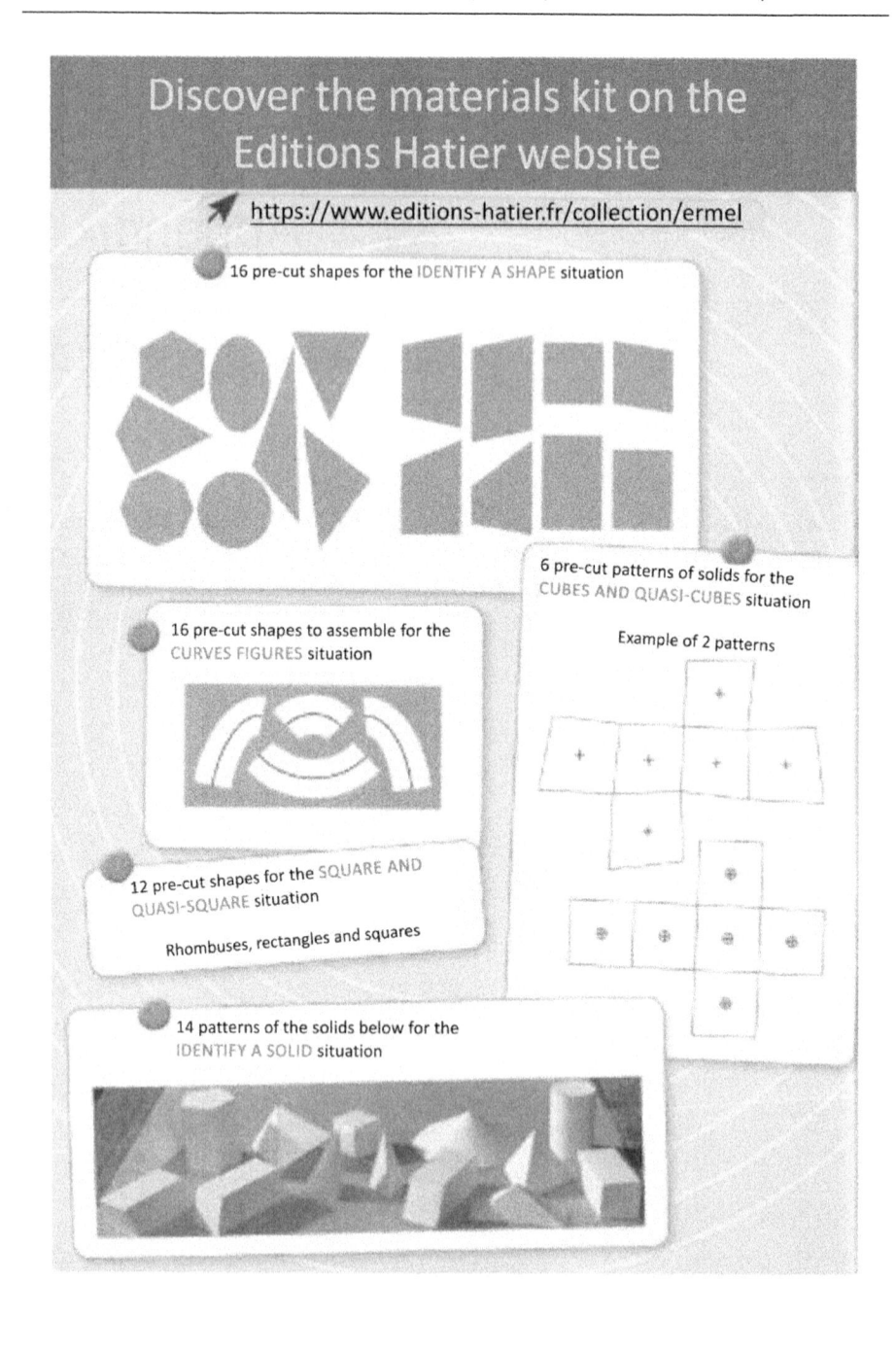

312 Articulations between Tangible Space, Graphical Space and Geometrical Space

Discover the materials kit on the Editions Hatier website

https://www.editions-hatier.fr/collection/ermel

16 pre-cut shapes for the IDENTIFY A SHAPE situation

16 pre-cut shapes to assemble for the CURVES FIGURES situation

6 pre-cut patterns of solids for the CUBES AND QUASI-CUBES situation

Example of 2 patterns

12 pre-cut shapes for the SQUARE AND QUASI-SQUARE situation

Rhombuses, rectangles and squares

14 patterns of the solids below for the IDENTIFY A SOLID situation

Appendix 4

Triangles within Quadrilateral

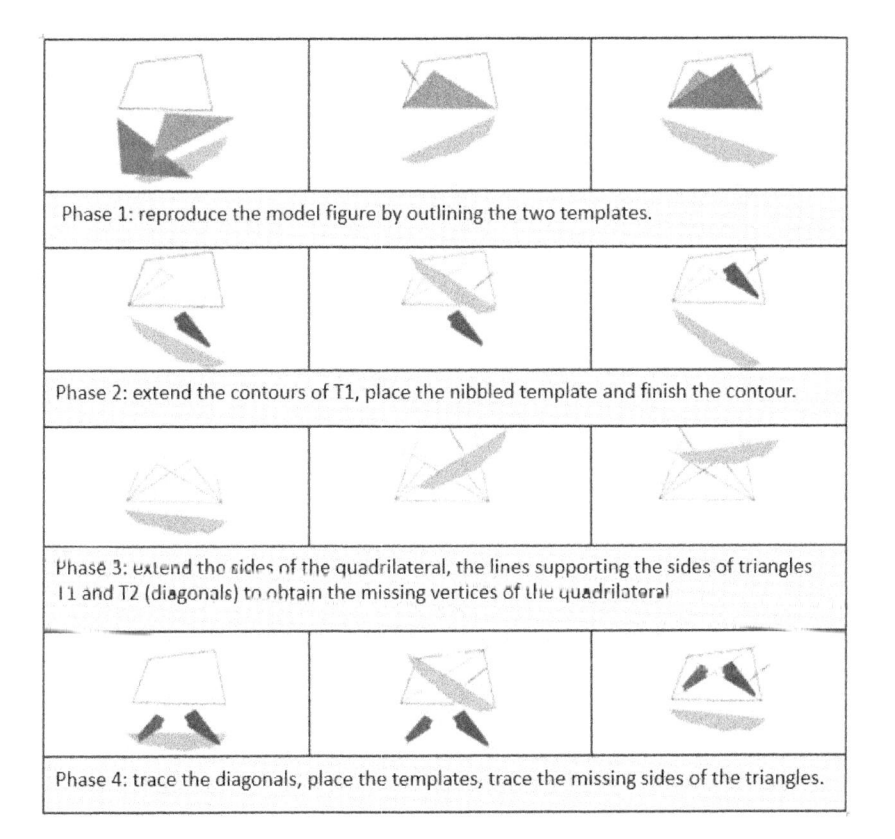

Phase 1: reproduce the model figure by outlining the two templates.

Phase 2: extend the contours of T1, place the nibbled template and finish the contour.

Phase 3: extend the sides of the quadrilateral, the lines supporting the sides of triangles T1 and T2 (diagonals) to obtain the missing vertices of the quadrilateral

Phase 4: trace the diagonals, place the templates, trace the missing sides of the triangles.

For a color version of this figure, see www.iste.co.uk/guille/tangible.zip

List of Authors

Henri-Claude ARGAUD
ERMEL
IFÉ
Lyon
France

Teresa ASSUDE
ADEF
Aix-Marseille Université
Marseille
France

Céline BEAUGRAND
STL
Université de Lille
France

Caroline BISSON
Université de Sherbrooke
Canada

Sylvie BLANQUART
LAB-E3D
Université de Bordeaux
Mont-de-Marsan
France

Valentina CELI
LAB-E3D
Université de Bordeaux
Pau
France

Sylvia COUTAT
Faculté de psychologie et des sciences de l'éducation
Université de Genève
Geneva
Switzerland

Jacques DOUAIRE
ERMEL
IFÉ
Paris
France

Fabien EMPRIN
CEREP, ERMEL
IFÉ
Université de Reims Champagne-Ardenne
Châlons-en-Champagne
France

Claire GUILLE-BIEL WINDER
ADEF
Aix-Marseille Université
Aix-en-Provence
France

Christophe HACHE
LDAR
Université de Paris
France

Catherine HOUDEMENT
LDAR
Université de Rouen Normandie
France

Christine MANGIANTE-ORSOLA
LML
Université de Lille
Arras
France

Patricia MARCHAND
Université de Sherbrooke
Canada

Emilie MARI
ADEF
Aix-Marseille Université
Marseille
France

Anne-Cécile MATHÉ
ACTÉ
Université Clermont Auvergne
Clermont-Ferrand
France

Catherine MENDONÇA DIAS
DILTEC
Sorbonne Nouvelle
Paris
France

Ismaïl MILI
Haute École Pédagogique du Valais
Saint-Maurice
Switzerland

Karine MILLON-FAURÉ
ADEF
Aix-Marseille Université
Marseille
France

Marie-Jeanne PERRIN-GLORIAN
LDAR
Université d'Artois
Paris
France

Edith PETITFOUR
LDAR
Université de Rouen Normandie
France

Céline VENDEIRA-MARÉCHAL
DiMaGe
Université de Genève
Geneva
Switzerland

Karine VIÈQUE
LDAR
Université de Lille
France

Index

A, C

action situation, 5, 9, 15, 22, 136, 224, 226
artifact, 48, 70, 96, 99, 103, 142, 150, 151, 224, 297
coding, 51, 73–79, 125, 140, 188, 211, 279, 282, 283
conceptualization, 6, 11, 17, 18, 21, 35, 37, 38, 40, 43, 44, 73, 136, 218, 247

D, E

didactic
 engineering, 117, 201, 219
 knowledge, 213, 217, 218
 quality, 181, 182, 191, 192
 repression, 236–238, 278
 reticence, 232, 237, 238
 variable, 36, 39–41, 174, 175
dimensional deconstruction, 4, 6, 8, 53, 68, 102, 114, 278, 284
dyadic work arrangement, 132, 135, 137, 138, 140, 142, 145, 146
dynamic geometry environment/dynamic geometry software, 77, 95, 96, 104, 134, 158, 279, 285, 297
dyspraxia, 130, 142, 150, 152

educational robotics, 126, 128
external transposition, 265, 266, 267

F, G

figure
 reproduction, 14, 15
 restoration, 5, 9, 12–14, 17, 19, 22, 27, 31, 35–38, 44, 297
formulation situation, 19, 23, 24, 26, 27
freehand drawing, 16, 81–83, 85, 87, 88, 89, 279, 282
geometric
 knowledge, 4, 47, 63, 70, 84, 87, 89, 134, 137, 176, 200, 208, 209, 213, 214, 217, 218, 242, 290, 297
 language, 14, 16, 17, 19–23, 31, 136, 145, 151, 188, 215, 290, 294, 296
 lexicon, 58, 75, 234, 277, 278
 shape, 61, 108
geometric thinking
 abstraction levels, 49
geometrical use of instruments, 12, 31, 37, 44
Geometry 1 (G1), 222, 292–296
Geometry 2 (G2), 222, 292, 293, 295, 296

geometry of tracing, 3, 5, 8–10, 101, 291, 294–296
gestures, 16, 54, 59, 62–64, 66–70, 88, 102, 104–106, 108, 130, 135, 136, 162, 167, 169, 173, 208, 209, 213, 238, 265, 278, 299
graphical
　expression, 114, 115, 281–285
　space, 73, 75, 101, 105, 107–109, 113, 114, 156, 157, 180, 277–282

H, I

haptic
　modality, 51, 53
　perception, 51–55, 57, 62, 69
human interaction simulator, 129, 130, 138, 141
institutionalization, 19, 165, 186, 210, 211, 213, 217, 222, 228, 236, 247, 269, 298
instrumental integration mode, 99
instrumented action, 33, 118, 135, 136, 140, 153, 176, 179, 194, 290
internal transposition, 267, 268

M, O

macro-space, 223, 246
mereological modifications, 49, 62
meso-space, 101, 104, 107, 121, 126, 156–158, 223, 225, 226, 228, 246, 297
micro-space, 104, 107, 121, 122, 126, 156, 223, 225, 226, 246, 250
old/new dialectic, 100
ostensives, 10, 98, 173, 180, 182, 186, 191, 270, 272, 277

P, R

parallelism, 74, 85, 87, 89, 138, 156, 172, 174–181, 183–185, 187–190, 192, 193
perceptual control, 87, 145

perpendicularity, 21, 22, 26, 74, 78, 140, 156, 172, 174–180, 183, 185, 187–190, 192, 193, 295
practices
　coherence of, 208, 218
praxeological integration mode, 99, 100
praxis continuity, 294
problem-solving, 12, 18, 89, 100, 156, 162, 167, 169, 183–185, 193, 242
production of resources, 170, 219
proof procedures, 17, 27, 31, 285
reasoning, 3, 84, 86–89, 120, 130, 145, 221, 222, 224, 292

S, T, V

scheme, 38, 41
semiotic
　potential, 48, 60, 70
　representation, 37, 114, 277, 283
sequential understanding, 52, 53, 57, 62
solids, 156, 157, 162, 165–169, 189, 237, 297, 311
spatial knowledge, 84, 87, 89, 101, 102, 119–123, 126, 178, 184, 185, 241, 271, 290, 297
straight line, 18, 41, 55, 145, 157, 158, 161, 234
Tangram, 247, 249, 250, 252, 254, 255, 257, 258
teacher training, 31, 97, 265, 290, 301
teacher's discourse, 210, 231, 232, 234
teaching practices
　enrichment of, 201
technical language, 16, 31, 110, 135–137, 151
textbook, 171, 181, 182
time economy, 97, 98
topological understanding, 50
validation situation, 29
visual perception, 48–50, 52–55, 57, 62, 66, 71, 83, 135, 162, 175

Other titles from

iSTE

in

Innovations in Learning Sciences

2022

BISAULT Joël, LE BOURGEOIS Roselyne, THÉMINES Jean-François,
LE MENTEC Mickaël, CHAUVET-CHANOINE Céline
Objects to Learn About and Objects for Learning 1: Which Teaching Practices for Which Issues?
(Education Set – Volume 10)

BISAULT Joël, LE BOURGEOIS Roselyne, THÉMINES Jean-François,
LE MENTEC Mickaël, CHAUVET-CHANOINE Céline
Objects to Learn About and Objects for Learning 2: Which Teaching Practices for Which Issues?
(Education Set – Volume 11)

DAVERNE-BAILLY Carole, WITTORSKI Richard
Research Methodology in Education and Training: Postures, Practices and Forms
(Education Set – Volume 12)

HAGÈGE Hélène
Secular Mediation-Based Ethics of Responsibility (MBER) Program: Wise Intentions, Consciousness and Reflexivities
(Education Set – Volume 13)

2021

BUZNIC-BOURGEACQ Pablo
Devolution and Autonomy in Education
(Education Set – Volume 9)

SLIMANI Melki
Towards a Political Education Through Environmental Issues
(Education Set – Volume 8)

2020

BOUISSOU-BÉNAVAIL Christine
Educational Studies in the Light of the Feminine: Empowerment and Transformation
(Education Set – Volume 6)

CHAMPOLLION Pierre
Territorialization of Education: Trend or Necessity
(Education Set – Volume 5)

PÉLISSIER Chrysta
Support in Education

2019

BRIANÇON Muriel
The Meaning of Otherness in Education: Stakes, Forms, Process, Thoughts and Transfers
(Education Set – Volume 3)

HAGÈGE Hélène
Education for Responsibility
(Education Set – Volume 4)

RINAUDO Jean-Luc
Telepresence in Training

2018

BARTHES Angela, CHAMPOLLION Pierre, ALPE Yves
*Evolutions of the Complex Relationship Between Education and Territories
(Education Set – Volume 1)*

LARINI Michel, BARTHES Angela
*Quantitative and Statistical Data in Education:
From Data Collection to Data Processing
(Education Set – Volume 2)*

2015

POMEROL Jean-Charles, EPELBOIN Yves, THOURY Claire
MOOCs: Design, Use and Business Models

Printed by BoD"in Norderstedt, Germany